Mathematicians as Enquirers
Learning about Learning Mathematics

Mathematics Education Library

VOLUME 34

The titles published in this series are listed at the end of this volume.

Mathematicians as Enquirers
Learning about Learning Mathematics

by

Leone Burton
The University of Birmingham, United Kingdom

and

In collaboration for Chapter 6 with
Nathalie Sinclair
Michigan State University, United States of America

KLUWER ACADEMIC PUBLISHERS
Boston / Dordrecht / New York / London

Distributors for North, Central and South America:
Kluwer Academic Publishers
101 Philip Drive
Assinippi Park
Norwell, Massachusetts 02061 USA
Telephone (781) 871-6600
Fax (781) 681-9045
E-Mail: kluwer@wkap.com

Distributors for all other countries:
Kluwer Academic Publishers Group
Post Office Box 322
3300 AH Dordrecht, THE NETHERLANDS
Telephone 31 786 576 000
Fax 31 786 576 254
E-Mail: services@wkap.nl

Electronic Services <http://www.wkap.nl>

Library of Congress Cataloging-in-Publication Data

Burton, Leone, 1936-
Mathematicians as enquirers: learning about learning mathematics / by Leone Burton and in collaboration for chapter 6 with Nathalie Sinclair.
p.cm. – (Mathematics education library; v. 34)
Includes bibliographical references and index.
ISBN 1-4020-7853-6 (alk. paper) – ISBN 1-4020-7859-5 (pbk. : alk. paper)
ISBN 1-4020-7908-7 (e-book)
1.Mathematics—Study and teaching. I. Sinclair, Nathalie. II. Title. III. Series.
QA11.2.B87 2004
510'.71 – dc22

2004044224

Printed on acid-free paper.

Printed in the United States of America.

DEDICATION

To my family:

Mark, Helen, Samuel (in order of height), partners in loving and learning,
together with the twins to come

and in loving memory of Simon Rosenblat, mathematician and dear friend

CONTENTS

LIST OF TABLES

PREFACE

It is amazing that the usual reply to being introduced to a mathematician is a stumbling apology about how bad someone is at mathematics, no matter how good they may be in reality. The problem is that we have come to view mathematics as an arcane branch of knowledge that only a few can aspire to understand or grasp. The sense of separation between those who have the knowledge and those who do not, is present even amongst academics where many of the same skills and research practices exist – intuition, the use of symbolic structures and the use of intuition and insight. The more worrying aspect of this separation is the ever declining numbers of students choosing mathematics as part of their curriculum beyond the stage when it is mandatory. Even worse, it would seem that the mathematics community has created a discriminatory environment that deters many students from continuing beyond secondary and tertiary education. Exit interviews with graduates show a stark rejection of their previously chosen subject.

What about mathematicians themselves – how do they see themselves? Do male and female mathematicians come to know their mathematics in different ways, do they tackle different problems with varying results for their status in the research community? Does the pervasive labelling that mathematics is a male domain arise because of a lack of role models for women or through its highly competitive, hierarchical nature? Are the negative labels often associated with mathematics, i.e. intimidation, elitism, reification and inaccessibility, the result of inordinate influence exerted by a non-representative group at higher levels?

Leone Burton provides us with answers to these difficult questions through a subtle blend of academic enquiry and personal insight in her study of a group of 35 pairs of mathematicians. Her quest – to find out how mathematicians come to know about their subject, the models they use, their experience in working in the mathematics community and their understanding of the differences of gaining knowledge in their research versus being taught or in teaching.

As humans beings we constantly use our innate learning skills to generate highly sophisticated mathematical solutions to life. Beginning with pattern association and language development, the vast majority of us has at one time or another experienced the satisfaction and thrill of being able to solve a puzzle or meet a physical challenge with just the right balanced movement. We know when something is right, intuitively and in reality. Taken to its limit, this means that anyone should be able to understand the fundamental basis upon which mathematics is built. Moreover, we do all this in a social context with it attendant aesthetics and cultural attributes. The epistemological model that Professor Burton uses in her studies enshrines these ideas by conjecturing that coming to know mathematics is a product of people and society. The model examines the extent of analytic, conceptual and visual thinking styles and the inter-dependence of mathematics on aesthetics and intuition. Her results are in sharp contrast to the Platonist view that suggests that mathematics is just waiting "out there" to be discovered, objective and independent of human values.

The mathematicians told of

> *five seconds, these moments when you feel you have seen one of the most beautiful things in the world;*
>
> *String Theory is something I can write down in about two lines and it looks very beautiful to me. It is true that I cannot explain its consequences without doing unpleasant things;*
>
> *there are nice proofs and nasty proofs, ... every so often you have to do hard analysis and it might be clever what people do but it looks hideous on the page;*
>
> *Lighthill's papers were such elegant pieces of mathematics. So many problems nowadays are computerized in a way that, for me, some of the excitement has gone;*
>
> *I suppose one gets a sense of exhilaration, an appreciation of its elegance – a fit of form and shape in a very abstract sense just as in certain paintings or music;*
>
> *I find a level of beauty in wooden puzzles. If things fit together nicely, in a smooth way, no gaps and it feels nice ... it is that sort of idea. A number of ideas fit together nicely with no gaps.*

Mathematics then appears to be about aesthetics, coherence, symmetry and function.

What about intuition? Some of the mathematicians were uncomfortable about saying they used intuition, but it was clear that something along the same lines played a key role in all their thinking. Intuition was interchanged with gut feelings, seeing the best way forward, getting *Ahas*! in the morning, or seeing short cuts to solutions. Intuition was the result of experience and knowledge, often unreliable and needed checking, but nevertheless an important step in mathematical thinking and research.

So how does it feel to be a mathematician? Leone Burton manages to draw out the passion and excitement of her interviewees and convey the pleasure they derive from knowing about mathematics. She places in the readers' hands, the essential messiness of the mathematician's arrival at 'truth' and asks us to think again at how we might transform the pedagogical style of transmissive teaching and inert knowledge in the classroom into one driven by mathematical enquiry.

The compelling personal stories of the mathematicians in the study put to rest the myth that mathematicians are born, into families with mathematical connections, rather than made. Almost all remarked upon the critical role played by a teacher in directing them towards a career in mathematics. The stories go a long way to underline the heterogeneous backgrounds and aptitudes of the group in the study. As to the problems they work on and their reasons for collaborating with others, these are equally diverse and demonstrate a considerable shift from half a century ago. For example, only 4 claimed to do all their work on their own; it was said by one that *there was bad atmosphere in research mathematics across the world, of the best mathematicians trying to show they were best... co-operation is important now.* The reasons for the shift were many, and underline the very essence of team-work in all parts of science and technology, that of being less isolated, sharing the euphoria of discovery, increasing the quality and quantity of ideas and increasing the likelihood of examining new ways of solving a problem.

Curiously, despite all the bonhomie of working together, competition within the mathematics community has not gone away. As some of the interviews serve to underline, competition still manages to weave its destructive way through the community, exposing itself as the exercise of power, often to the detriment of publishing practices. Several women note that it is imperative to understand the power base of the community in order to see how knowledge is created, refined and acknowledged, so that they might be better accepted.

The hierarchy of judgement, as Leone Burton calls it, refers to the use of words such as significant, important, interesting and trivial. Such terms are used to label findings, but the fundamental question – should a paper contain correct mathematics? – was met with a more flexible view as to how much effort a referee should put in to confirm this! That a paper should contain novelty however was not in dispute.

So "was it different for you?" – it certainly seems from the interviewees that diversity is not an attribute welcomed within the mathematics community. Moreover, if you are a female mathematician you are likely to be the subject of sexist remarks, usually relating to limitations arising from being emotional, dependent, not quite up to it, or not being highly focused. This of course denies the human dimension that quite clearly springs up from the discourse and interviews described in earlier chapters by both men and women. But more worrying are the negative comments arising from the presence of hierarchies in which women often distinguish themselves from "real" mathematicians, accepting that the rules of the promotional game also appear to be different for women. Sobering comments emerge in this regard from the women participants who had been at Oxbridge during their training.

Hierarchies institutionalize competition in universities and schools, so that right from the beginning *people are made to feel stupid if they don't achieve in mathematics*. It is certainly not a field in which the students should expect to be nurtured. Oh, and by the way, giving seminars in which the mathematics looks too easy is clearly not the way to achieve membership in learned societies or honours.

Learning mathematics, we are told by Leone Burton, needs mathinking. Unfortunately, there is an air of arrogance, smugness, intimidation, pressure, isolation and competition that needs to be overcome by the mathematics community itself before any progress can be made. That it can happen is without doubt possible; everyone interviewed fell quickly into the joy and euphoria of discovery and knowledge when talking about their research. But to find out just how to do it, and the levers that will need to be pushed, you will need to read the book. It is riveting and realistic, and a hard reminder of just what we are losing when we turn off young people from experiencing the dizzying heights of mathematical enquiry.

Jacqueline McGlade
Executive Director, European Environment Agency, Copenhagen
& Professor, Department of Mathematics, University College London

ACKNOWLEDGEMENTS

All seventy participants in this study deserve my heartfelt thanks for their time, their willingness to share sometimes very personal histories or experiences and for helping to create a bridge, even if not yet very robust, between mathematics and mathematics education. In the same breath, I would like to thank Professor Jacqueline McGlade for giving her time, and interest, to this project in the form of the Preface. In addition, sincere thanks go to Hannah Bartholomew, Marion Bowl, Bob Burn, Christine Hockings, Tony Ralston and members of the King's Mathematics Education Study Group for their critical reading skills which, I am sure, have vastly improved the end-product but for which they are in no way responsible. For helpful reading, advice and, always, interpersonal support, thank you Mary Barnes. And for technical and production support, thank you Julie Foster. The writing of Chapter 6 with Nathalie Sinclair was a pleasure through which we have formed a friendship that I value. All of my research students have, and continue, to contribute to my thinking in enumerable ways and I am grateful for the rites of passage that we have shared. This comment also applies to friends, sometimes research students of others, who will, I hope, recognise themselves and their contributions.

Leone Burton
London, UK, December 2003.

1

Why this book?

In principle, this book is the result of a study of the research practices of mathematicians. The study was an attempt to get inside how mathematicians construct their own knowing of their discipline through their research. But, in reality, undertaking that study was my response to a professional lifetime of trying to understand what I perceived as an irreconcilable gap between how I heard mathematicians describing the pleasures and pain of their craft, and how I experienced, and observed, them teaching mathematics. Research has confirmed that mathematics in higher education is content-dependent and didactically traditional (see, for example, Burton & Haines, 1997).

The power of mathematicians to influence the learning of mathematics cannot be underestimated. Under the present arrangements for educating secondary teachers in the UK, for example, mathematicians are central. This is not only because all teachers must pass through their hands to become mathematics graduates but also because the transmissive teaching strategies that many exemplify often have been the sole experience of the learner from the age of 5 until graduation. This continuity of receptive learning may sometimes appear to be effective when measured in a formulaic way; but it distances most learners from experiencing mathematics as a creative, exciting and challenging undertaking based on their own enquiries.

I remember, when first faced with the need to solve a problem (*not* an exercise), my own shock at how little my successful learning of mathematics so far had prepared me. On top of that shock, I recall also the amazement at how difficult, engaging, and ultimately satisfying was the experience. This left me with two professional questions. Why are learning opportunities to meet and engage with mathematics so limited? Whose interests are served by the preservation of the gap between codified, transmitted mathematics and mathematical enquiry, whether that is called problem-based learning, problem solving, investigations, 'realistic' mathematics education or so-called authentic mathematical assessment? Not that all of these are identical each to the other. But they share an approach to the learning of mathematics that rejects the pedagogically inadequate idea that it is possible to decide for every learner exactly what they will learn, when and how. Or that it is possible to measure that learning through limited, timed, written tests.

I, too, have been the object of that cocktail-party type conversation where the answer to the question "What do you do?" is greeted by "I was never any good at mathematics". But such experiences, which all mathematics insiders can claim, are reinforced by the results of many different research studies which emphasise how uniform, and disagreeable for some, is the learning experience of mathematics. Examination results and the serious decline in numbers, internationally, of those choosing to study mathematics confirm this. As a result of her study of the teaching of mathematics in two very different schools, Jo Boaler pointed to "the potential of open and closed teaching approaches for the development of different forms of knowledge" (1997a: 152). Furthermore, a shift from 'project-based' work back to more formal mathematics, to 'prepare' for public examinations, in her study induced "a corresponding change in the students' views about mathematics" (1997a: 61), a change which was also recorded by Paul Cobb and his colleagues (Cobb et al., 1992). Jo Boaler explained the breadth of her study:

> The idea that traditional models of teaching disadvantaged students did not come from my own predispositions or from individual students' descriptions. It came from the repeated reports of students, which were supported by observations in lessons, questionnaires and various different assessments. (Boaler, 1997a: 151)

I also undertook some studies to introduce mathematics learners to an 'enquiry-based approach' and found, very rapidly, how productive this could be for their learning and their attitudes. Despite such research results, however, many learners continue to be offered a very limited diet and experience in mathematics.

The form of knowledge promoted by transmissive teaching is what A.N. Whitehead (1962) called "inert". It is passed, like a parcel, from the knowing teacher or textbook to the learner, a practice found particularly in the mathematics classroom where research has established that there is a dependency upon textual materials not found in other subject areas. (See, for example, Burton, 1994, for a comparison of the epistemological base for the teaching of mathematics and of English.) This pedagogic practice has, implicitly within it, forms of control which situate the learner as naïve and unknowing and, consequently, as dependent upon the teacher to gain what the teacher, the school, the society, have designated as knowledge necessary to be successful. The learner is controlled, both epistemologically by how knowledge is understood, and pedagogically, by how learning is experienced, since these are presented as a package. The teacher can manipulate pace and pressure with the result that only particular students survive – those who can adapt to, or are comfortable with, these conditions. The hallmark of their survival is usually marked by achievement on public examinations.

WHO ACHIEVES?

Observations such as these have led many researchers to point to pedagogical factors as causes for declining interest in studying mathematics. Those interested in why fewer women remain in mathematics, the further 'up' the educational system one looks, have been particularly outspoken about the effects of classroom environment, for example competition, on student interest (see, for example, Boaler, 1997b, 2002, Herzig, forthcoming).

Despite the hysteria around 'What about the boys?' it is noticeable that the proportion of women, out of the diminishing number of students who present themselves to study mathematics at each level where there is a moment of choice, decreases. Abbe Herzig (2004) pointed to the following attrition rate in the USA:

Bachelor's degrees in mathematics awarded to women in 1994	47%
Master's degrees in mathematics awarded to women in 1996	40%
Doctoral degrees in mathematics awarded to women in 2000	27%
Full-time female staff employed to teach mathematics in 2000	22%

She commented "While these numbers represent substantial improvement over recent years, they indicate that the participation of women in post-secondary mathematics remains problematic" (forthcoming). In a paper presented at a one-day conference on the shortage of mathematics teachers in UK secondary schools, I also drew attention to the overall decline in numbers of students choosing mathematics. I reported on the responses of 'Advanced Level' students, aged 17, who had opted to study mathematics, to the question of whether they would undertake any further study of mathematics; 73.3% unequivocally said no, including 84.6% of the female students (Burton, 2001a: 63). And a report in The Times Higher Education Supplement of September 28, 2001 said:

> Australia will have no mathematicians working in its universities by 2012 if trends continue…The number of secondary-school students taking courses in physics, advanced mathematics and chemistry over the past 20 years has also fallen. The academy [of Technological Sciences and Engineering] says that if this decline continues, these subjects will disappear from the school curriculum by 2020. (p.12)

In The Guardian of March 4, 2002, it was noted, for the UK, that:

> Statistics collected by the London Mathematical Society bear out the story that at every stage beyond school, increasing numbers of women drop out of maths. In 1998 (the latest available figures) women comprised 38% of maths undergraduates. At postgraduate level that proportion had fallen to 29%, while just 18% of university lectureships were held by women. Women made up 7% of senior lecturers and 2% - nine individuals - of university professors.

Women are, perhaps, the more visible sector of what is an international trend of declining numbers engaging with mathematics. It has long been my opinion that the mathematics experienced by students in formal education, and the ways in which it is encountered, offer explanations for this decline in interest. Public interest books about mathematics are readily bought so it cannot be that people have no wish to engage with mathematics. Zelda Isaacson, writing of a return-to-study mathematics course, quoted a female student who said:

> Especially in maths, I couldn't stand the competitiveness of seeing the 'clever boy' (there's always one in every class!) do the exercises miles before everyone else and understanding things quicker than everybody which just reinforced the idea that I didn't have a 'mathematical brain' (1990:23).

Despite such experiences, this student had made a choice to re-enter science and technology, which involved engaging with mathematics; a similar choice is made by many mature men and women sometimes in an attempt to wipe out what they have seen as earlier personal failures (and see Bibby, 2002).

In a number of studies, along with other researchers, I have shown that once students make a choice to study mathematics, many of them report experiences that are not conducive to holding them in the discipline. The same pattern holds whether they are studying mathematics at school, as a pre-university choice, at university as undergraduates or even at doctoral level. This rejection of the discipline as experienced by learners, is not only a phenomenon of those who drop-out but can also be found amongst those for whom the study of mathematics has gained them a qualification. Out of 31 successful graduates from a highly prestigious UK university mathematics department, 54% of the females and 25% of the males answered my question about what they planned to do next along the lines of 'whatever it is, it won't be mathematics'. One student commented[1]:

> *The way in which [this university] carefully selects the best candidates for each subject and then fails to fully exploit them is bordering on the criminal. For the sake of snobbery and tradition, exciting, effective modern teaching techniques are not used, and lectures are given in unnecessary abstraction.*

In an American study of university student attrition rate from science, mathematics and engineering studies, Elaine Seymour and Nancy Hewitt stated:

> The loss of 40 to 60 percent of a group of students with higher than average abilities within two years of taking their first college science or mathematics class would be a serious matter notwithstanding the evidence of higher loss rates among those minority students (by gender, race or ethnicity) on whose recruitment and retention much money and effort has been expended (1997: 391).

They continued:

> It is also clear from our data that the most effective way to improve retention among women and students of color, and to build their numbers over the longer-term, is to improve the quality of the learning experience for all students (Ibid: 394).

Abbe Herzig's study of doctoral students in mathematics, where she quoted an attrition rate of between 30% and 70% demonstrating further disaffection, identified a need for interaction, participation and integration as crucial factors in student retention (Herzig, 2002b). These students identified the absence of such factors in their research environment.

Students such as those mentioned above, and many others encountered in these pages, complained about their pedagogical experiences as well as the environment in which they learned. However, staff continue to believe that students must learn 'the basics' before embarking upon anything more challenging and, furthermore, that "mathematics is a very difficult field of study, and that only some people have the talent required to be successful at mathematics" (Herzig, 2002b: 184).

---

[1] Throughout this book, direct quotations taken from transcriptions of research data are shown in italics.

THE STUDY AND THE BOOK

I undertook the study reported in this book in the belief that these attitudes and behaviours could, in part, be traced back to the epistemological perspectives of the mathematicians teaching the students. The beliefs in mathematical difficulty and in special inheritance of 'ability' (for a discussion of what makes 'ability' so problematic, see Lynch & Lodge, 2002) are closely connected to an hierarchical view of mathematics which identifies 'the basics' and then leads to a transmissive style of pedagogy. This reflects a position on knowledge rather than knowing. It reifies the objects of mathematics and removes the person from their exploration or acquisition. If this is a good description of how mathematics is experienced at university (and it is a description borne out in many different research studies), it is certainly a good description of the ways in which many schoolteachers, as well as members of society more generally, understand mathematics and how it should be taught and learned. But is this the way that mathematicians themselves learn when they are researching? I doubted that it could be so and in this book I provide the evidence that their stance on how *they* come to know when researching is very different from the one that drives the teaching of their students. I make no apology for constructing a general argument about learning on the basis of interviews with practising mathematicians given the advice of Jerome Bruner:

> if one wishes to study the psychology of mathematics (as, say, G. Polya did), one studies the works of trained and gifted mathematicians, with particular emphasis on the heuristics and the formalisms they use to give form to their mathematical intuitions (1986: 15).

I deliberately set out to interview equal numbers of female and male mathematicians so that I had grounds for making comparisons between their beliefs, attitudes and experiences. However, this is not a book about gender issues. There are general features that, I believe, are helpfully explained by using gender as a way in, a perspective, that can open out these issues to a wider consideration. I believe that female/male differences, where these exist, can tell us something about the climate and the experiences of all learners so, in all but Chapter 9, gender is a lens, not an issue. However, as I explain below, Chapter 9 specifically focuses on the differences for females and males in their experiences of the mathematical culture.

Before outlining the remainder of the book, I would like to offer the reader a word of warning. This book describes the ways in which mathematicians come to know, learn, hence my lack of apology about generalising from their experiences to other learners of mathematics, even though the level of learning in terms of mathematical sophistication may be very different. For this reason, I freely draw on examples from the mathematics education literature about learning of young children or adolescents in schools as well as students at universities. The focus of this book is on what, from mathematicians *as* learners, we can learn about learning mathematics; it is not on learning at a particular age or stage.

In Chapter 2, I outline the model that I devised to describe mathematical knowing and then explain how the study was set up to investigate how good a match was the model to the ways in which the mathematicians described their practices. Chapter 3, while preserving the anonymity of the mathematicians, provides information to help contextualise them; their backgrounds, the range of universities where they worked, their employment status and other factors which emerged from the interviews are all included.

The focus of attention in Chapters 4, 5 and 6 is the epistemological model. In Chapter 4, I explore one of the surprising outcomes of the study, that, when thinking about the problems they researched, the mathematicians described that thinking in three different ways. This was surprising because only two styles of thinking about mathematics had been recorded in the literature. Furthermore, only three of the mathematicians used all three styles of thinking, others relying upon only one, or two.

Aesthetics and the role of intuition or insight are much discussed in the mathematics literature and hence formed two of the categories in the model. In Chapter 5, I describe how these mathematicians spoke about aesthetics and intuition/insight and I explore the feelings that they expressed about mathematics and its practices. We meet some of the mathematicians in Chapter 6, written jointly with Nathalie Sinclair. We use a case study approach to the personal context of the individual mathematicians to provide further detail on the ways in which the categories of the model were discussed.

In Chapter 7, I look at the practices of the mathematicians when researching and, in particular, discuss their use of collaboration and co-operation. They explain how they find and choose the problems on which they work and which are of interest to them, again raising issues of collaboration and networking within their community. An important part of the interview was devoted to discussing the outputs of their research through publication, what should be published and why. These are presented in Chapter 8. I set out on the study with expectations that, between the female and the male mathematicians, I would find no epistemological differences. However, in Chapter 9, I explore the differences that I *did* find, differences that relate to the distinctive mathematical culture (that is, the socially organised features of the community) rather than the culture of mathematics (that is the structure 'imposed' by the discipline). This distinction is met many times in the book. In the process of being socialised into these cultures, female mathematicians apparently offer both a challenge and an impetus for change.

Out of the study came a lot of support for the claim that the practice of mathematics is not the cognitively dominant and 'objectively' clean discipline widely presented and believed. Because the pursuit of mathematics is a human, socio-cultural activity, talking about it invokes feelings that are expressed, in direct quotations, throughout the book. Feelings and cognitions interact for the mathematicians as they come to know. Of course, the implications for how the subject is taught and learned are great and important for all, mathematicians and non-mathematicians alike. These implications are explored in Chapter 10 where the epistemological base confirmed by the study is recognised for the role it plays in the practice of mathematicians and in society generally. If mathematicians are coming to know, then they are learning. What can we learn from their learning which is appropriate to apply to less sophisticated learners of mathematicians in schools and universities? A challenge is made to current thinking about teaching and learning of mathematics from this perspective; it is my aim that mathematics should be recognised as a part of everybody's entitlement.

The mathematical culture, explained in the practices described in Chapter 9, leads to discriminatory experiences that are enough to persuade some students that mathematics is not for them. Many of these practices are described throughout this book and, indeed, in research studies that I, and others, have done and to which reference is also made in these pages. A flavour of the fear and resistance of students can be found in messages on internet networks where questions such as the following have recently appeared:

> Can you recommend a school that either has a female professor or a male professor that would encourage/back a female applicant?
>
> Can you recommend a school that has a reputation for treating women applicants/grad
> students well? Or that is interested in increasing the number of female grad school students in their school?

One respondent to such a question wrote:

> There are quite a few departments in the world which have rude faculty members who either make sexist or sexual remarks. So you are smart to ask about departments before applying. In the United States this behavior is illegal, however, very few women prosecute so it still occurs.

Dr. Helen Robinson, a mathematics lecturer, in a report in The Guardian (4.3.2002) recalled her experience as the only female in a 45-strong postgraduate cohort at Warwick University 25 years before:

> When I was doing my MSc, if I gave the right answer to a question in a lecture and some male students had got it wrong before me, the lecturer would say: 'Look she can do it, and she's only a girl'.

Dr. Susan Howson, the first female in 120 years to win the prestigious Adams Prize for mathematics, in the same report expressed the belief that women are still deterred from working in pure mathematics not only by the lack of role models - she had only male teachers as a Cambridge undergraduate - but also by its highly competitive nature.

Many of the mathematicians who participated in this study spoke about competition as a destructive feature of the mathematical culture but competition was not the only thing about which they complained. In mapping their descriptions of how, with whom and under what conditions they research, I believe that I am outlining a culture, together with its epistemology, which is explanation enough for the declining interest in mathematics evidenced by possible students. But, the many contradictions and paradoxes to be found are also, I believe, cause for optimism. If we can release the stultifying control that is exercised over mathematics by an object-oriented epistemology, a concentration on knowledge, rather than on knowing, it is my hope that the delight, thrill, excitement, indeed to use the mathematicians' word – euphoria – to be gained from a study of mathematics, could become available to all learners. That is my aim in writing this book.

2

A model of how mathematicians come to know – creating, theorising, researching

In this chapter, I begin by explaining why I felt a need to create a model for how mathematicians come to know and how that model was derived out of the theoretical literature. I then address four epistemological challenges to mathematics arising from objectivity, homogeneity, impersonality and incoherence as all four create particular conditions for learning. In addressing these challenges, I explain how the categories of the model were derived as a consequence of a socio-cultural perspective on mathematics and its learning. In the final part of the chapter I provide information on the empirical study. I explain how, to test the epistemological model, I sought the participants, how the study was designed and the data were analysed.

CREATING AND JUSTIFYING THE MODEL

In Burton (1995), I described an epistemological model of mathematics built from the philosophical and sociological literature. The model presented the process of coming to know in mathematics in terms of five interacting categories:

* person- and cultural/social-relatedness,
* the aesthetics it invokes,
* its nurturing of intuition and insight,
* its recognition of different approaches, and

* its connectivities.

Why derive a model of coming to know mathematics?

There were three main purposes for deriving an epistemological model. First, I wanted to know if differential interest and performance in mathematics, for example between females and males, could be explained epistemologically, as well as pedagogically and in terms of curriculum and assessment. An extensive literature existed on the treatment of girls and boys in classrooms, as well as on the biased construction of textual and assessment material. This research was necessary and helpful but I had felt, for a considerable time, that mathematics itself, as a way of viewing and knowing the world, required epistemological investigation.

Second, I had been engaged in attempts to broaden the scope of mathematics learning in schools from the simple "delivery" of known facts to include learners actively undertaking their own enquiries as a necessary part of their learning. These enterprises had foundered on a widespread assumption in the UK, as, for example, articulated in *The Times Higher Education Supplement* by Professor David Crichton, then Head of the Applied Mathematics Department at the University of Cambridge. He said, "a common cause for concern is that there is too much emphasis on self-discovery rather than the presentation of material as a body of knowledge" (24/2/1995: 6). It was my interest to discover whether, in their own learning as researchers, mathematicians relied only on a 'body of knowledge' approach or invoked a set of processes, which I could describe in an epistemological model. If such a model could be verified, it would offer grounds, from within the discipline itself, for a shift in perspective away from a received knowledge view of teaching/learning mathematics.

Third, prior to an empirical investigation, I needed a tool, which would provide me with a theoretical description of how mathematicians come to know and, against which I could then map what they told me, in an empirical study. A model, which was justified out of the literature, seemed to me to provide a stronger basis for such an investigation than a description derived as "grounded theory" (Glaser & Strauss, 1967). To identify its categories, I had to examine an extensive literature base within mathematics, mathematics education, the sociology of knowledge of mathematics and science, feminist science and related areas. As a result of these enquiries, I derived the model outlined above (and see Burton, 1995).

I was conjecturing that coming to know mathematics is a product of people and societies, that it is hetero- not homogeneous, that it is inter-dependent with feelings especially those attached to aesthetics, that it is intuitive and that it inter-connects in networks. Additionally, the different approaches with which I was concerned included thinking styles. Two thinking styles had been recorded in the literature, a visual and an analytic. I conjectured that research mathematicians would use both and would move freely between them. If this model was validated in discussions with mathematicians, I felt that it would offer a high degree of support for arguments that present mathematics as a discipline with which to engage actively, instead of "an inert body of knowledge" (Whitehead, 1962: 2) to be transmitted and learned.

What was the theoretical base of the model?

The model I developed reflected considerable work that had been done in science. But it also extended ideas from the feminist science literature about the epistemological views which might constrain or facilitate enquiry and related these ideas to mathematics. In particular, I responded to feminist calls for an epistemology that was holistic, inclusive, accessible and heterogeneous (see, for example, Rogers, 1990, Ruddick, 1996). With respect to science, Sue Rosser (1990) said:

> If science is socially constructed, then attracting a more heterogeneous group of scientists would result in different questions being asked, approaches and experimental subjects used, and theories and conclusions drawn from the data (p.33).

I saw no grounds for believing that mathematics, in this respect, was different from science. Nor in the following respect:

> Since all descriptions of the world must pass through language, rhetoric, and metaphor, no description can claim it is free from the effects of human interpretative or rhetorical intervention...scientific knowledge cannot be separated from culturally derived representations or from the historical position

> of the interpreter doing the representing. It is as culturally formed as any other mode of discourse. (Ward, 1996: 41/2)

The work which I found convincing called for an integration of the knower with the knowns and

> Involves encouraging rebellious spirits to blossom with free rein to the imagination, preserving a certain nimbleness of mind while affording it the means of being creative. (Flato, 1992: 75)

Flato went on to compare such a regime with that found more conventionally:

> The 'training' procedures, as we conceive them and ordinarily practice them, hardly lend themselves, one must admit, to that kind of enticement, since they more often emphasize the transmission of acquired knowledge and apprenticeship in proven methods. And considering that those procedures resemble an obstacle course where the competition is tighter and tighter, this hardly encourages departing from the beaten path. (Ibid: 75)

The experiences of students of mathematics are, generally, well described by Flato. However, the conjecture that mathematicians as learners/researchers might be very different to how they are as teachers was supported by the writings of mathematicians such as Luke Hodgkin (1976) who discussed "an increasingly critical reflective viewpoint on the practice of mathematics itself" (p.30). He said:

> mathematics is not so much a hard crystalline mountain of truth painstakingly built up by successive generalizations as a field for error, for the conflict of ideas, for vagueness, for continual approximation and conjecture as the means of progress. So it's recognized among mathematicians, if not outside, that we do mathematics one way, but we might well do it another. (Ibid: 30)

And Joseph Weizenbaum (1976) added the warning that "all empirical science is an elaborate structure built on piles that are anchored not on bedrock as is commonly supposed, but on the shifting sand of fallible human judgment, conjecture and intuition"(p.14/15). In discussing the scientist's need to "believe his working hypothesis, together with its vast underlying structure of theories and assumptions, even if only for the sake of argument", he went on to say "I choose the word "argument" thoughtfully, for scientific demonstrations, even mathematical proofs, are fundamentally acts of persuasion" (Ibid: 15). This is certainly not the way in which mathematics is presented to learners in classrooms. I was struck by the differences between these perspectives on mathematics and those experienced by the majority of learners. I wondered how widespread amongst mathematicians were such views and about the implications of such differences for learning.

If the mathematicians' learning through their research is well described by my model, mathematics comes to be known in ways more closely related to an heterogeneous, inclusive, accessible and holistic epistemology. I saw such an epistemology as feminist because of:

* endeavour towards holism (Rogers, 1990; Ruddick, 1996);
* support for the person(s) claiming to know, within a community of practice (Wenger, 1998);
* concerns about the participatory practices of that community as well as about the manner in which the knowns are reified, valued and used. (Harding,1991; Henrion, 1997; Rose, 1994; Wertheim, 1997).

There is evidence (see Boaler, 1997a, 2002; Burton, 1990; Forgasz, 1994; Stein et al., 2000; Stobart et al, 1992) that learning mathematics at school level is more successful under a regime of this kind than taught traditionally. Within such an environment, there is an emphasis on the meaning which is made of the mathematical activities, the kinds of observations undertaken, the community practices of negotiation, conjecturing, testing, constructing argument, convincing and reflecting, all of which, amongst other reasons for valuing them, are part of the identified differences between a female- and a male-friendly science (see Damarin, 1991, Hekman, 1990, Rose, 1986, Rosser, 1990). But they also, interestingly, are consistent with the views of the mathematicians already quoted above.

The model was constructed from the now extensive literature which questions received science, most particularly its reductionism and claims to be objective and value-free (e.g. Harding, 1986, 1991; Keller, 1985, Knorr-Cetina, 1981, Rose & Rose, 1980). Even though Polanyi and Prosch could write in 1975 that "the ideal of pure objectivity in knowing and in science has been shown to be a myth" (p.63), that myth, to this day, continues to exercise considerable power over the curricular experiences of learners as, indeed, over the public presentation of mathematics by mathematicians (see Burton & Morgan, 2000,

Appendix A in this book). Unlike the public image of mathematics as being fixed and certain, Karin Knorr-Cetina wrote:

> whether a proposed knowledge claim is judged plausible or implausible, interesting, unbelievable or nonsensical, may depend upon *who* proposed the result, *where* the work was done, and *how* it was accomplished. (1981: 7, original emphasis)

Luke Hodgkin went even further:

> We have important ways in which scientific knowledge changes character, even locally, with changing culture. And this determining character of social institutions on the knowledge acquired is even *well-known* among scientists: X was at Princeton, at Grenoble, at Warwick *and so* he/she sees a differential equation in a particular way, even has a particular notion of 'proof'. But this knowledge is not spread outside the restricted community of scientists, so that the outsiders still believe in a universal agreement on an abstract 'rightness', which is something they themselves have been *taught* to believe in – at school, and in mathematics in particular. (1976: 39, original emphasis)

In addition, scientists have been challenged for their conventional pedagogical style. Enquiry methods, it is suggested, are often mechanistic and intrusive, separate the observer from the observed and reinforce competition (Kelly, 1987, Whyte et al., 1985). The style, translated into the classroom, presents students with received 'wisdom', and, in publications, the knowledge is offered as independent of the knowers who derived it, often with an existence of its own.

Although a much greater literature base existed in science than in mathematics, much of it the work of feminist scientists, a similar challenge has been offered to the epistemology of mathematics. In part, this emanated from the early work on gender differences in achievement in mathematics classrooms and, in part, has been provoked by a resistance to the marginalisation and stereotyping of women within the field. This movement has been supported by the work that has been done by historians of science, sociologists, philosophers, and anthropologists of scientific knowledge, particularly as their approach applies to mathematics. (See, for example, Bloor, 1991, Harding, 1986, 1991, Restivo, 1985, 1992, and Restivo et al., 1993).

Out of the literature, I identified four epistemological challenges to mathematics that have profound implications for how the subject is presented, experienced and understood especially by learners. These are challenges to objectivity, to homogeneity, to impersonality and to incoherence.

FOUR EPISTEMOLOGICAL CHALLENGES

From Objectivity to Mathematics as a Socio-Cultural Artefact

The major challenge was to a mathematics that is positivist, 'objective', value-free and independent of the human. Such a view reifies the 'knowns' and removes the knower. Many mathematicians, adopting a Platonist perspective on the discipline, believe that the mathematics is 'out there', awaiting their discovery. If they fail to turn over the right stone, another mathematician will do so. Such 'evidence' as is cited for this perspective is the compelling unity and consistency of the body of knowledge. But, the drive to consistency and coherence could be said to be an extremely human drive – it is, after all, what motivates very young children to learn. As human beings, we set up the conditions that we then 'find' to sustain our revelations. "If one works within a set of agreed-upon assumptions, it is likely that one's conclusions will share substantial common ground with those of colleagues also working within those assumptions" (Hunter, 1999: 135). But Brian Rotman (2000) pointed out that:

> It does look as if Platonism, if it is going to insist on timeless truth, is incapable of giving a coherent account of knowing, and a fortiori of how mathematical practice comes to create mathematical knowledge. (p.33)

I was convinced that "the linguistic system is part of the social system" (Halliday, 1975: 120) and equally convinced that mathematics was part of the linguistic system. Furthermore, as with language, I understood the acquisition of mathematical competency as two-way:

> The child builds up a potential for exchanging the meanings that are engendered by the system (and so, in the long run, for modifying the system-since the social system is a system of means, it is constituted out of innumerable acts of meaning which shape and determine the system). (Ibid: 121)

And I accepted that "the social context is therefore not so much an external condition on the learning of meanings as a generator of the meanings that are learnt" (Ibid: 140). The context of the situation within which the mathematics is placed produces 'talk' which "does not reflect their structure in any unprocessed or 'objective' way, but as it is processed by the culture" (Ibid: 141). As George Lakoff and Rafael Núñes expressed it:

> Human mathematics, the only kind of mathematics that human beings know, cannot be a subspecies of an abstract, transcendent mathematics. Instead, it appears that mathematics as we know it arises from the nature of our brains and our embodied experience. As a consequence, every part of the romance appears to be false. (2000: xvi)

The Romance of Mathematics was, they asserted, what "a great many of those who have a serious knowledge of mathematics not only tend to believe...but tend to believe it fiercely...part of their world view, their very identity"(p.339). They described the Romance in the following ways:

- Mathematics is an objective feature of the universe; mathematical objects are real; mathematical truth is universal, absolute, and certain.
- What human beings believe about mathematics therefore has no effect on what mathematics really is. Mathematics would be the same even if there were no human beings, or beings of any sort. Though mathematics is abstract and disembodied, it is real.
- Mathematicians are the ultimate scientists, discovering absolute truths not just about this physical universe but about any possible universe.
- Since logic itself can be formalized as mathematical logic, mathematics characterizes the very nature of rationality.
- Since rationality defines what is uniquely human, and since mathematics is the highest form of rationality, mathematical ability is the apex of human intellectual capacities. Mathematicians are therefore the ultimate experts on the nature of rationality itself.
- The mathematics of physics resides *in* physical phenomena themselves – there are ellipses in the elliptical orbits of the planets, fractals in the fractal shapes of leaves and branches, logarithms in the logarithmic spirals of snails. This means that "the book of nature is written in mathematics," which implies that the language of mathematics is the language of nature and that only those who know mathematics can truly understand nature.
- Mathematics is the queen of the sciences. It defines what precision is. The ability to make mathematical models and do mathematical calculations is what makes science what it is. As the highest science, mathematics applies to and takes precedence over all other sciences. Only mathematics itself can characterize the ultimate nature of mathematics. (p.339-340)

Nonetheless, the "Romance provides the standard folk theory of what mathematics is for our culture" (Ibid: 340) and has some extremely negative effects. George Lakoff and Rafael Núñes listed, among these, intimidation, elitism, reification and inaccessibility (see Ibid: 341). What I, and many others, have noted is that the results of these effects can be devastating for those who experience them, leading as they do to withdrawal from the discipline and self-blame. But, I believe, the results of these effects are also negative for mathematics because they are not equally distributed across the population so that those engaging with mathematics become less and less representative, the higher the level of study or work examined. Since it is my contention that mathematics is rooted in its socio-cultural context, a non-representative group of mathematicians must have inordinate power and influence on the discipline itself.

Perspectives on 'objectivity' in mathematics provoke related perspectives on the kind of person who best fits the mould of the 'objective' mathematician. There have been many observations about the paucity of female mathematicians (and mathematicians of colour in European-dominated societies) and one explanation that has been offered is that mathematicians are born, not made, to which Margaret Murray (2001) referred as "the myth of the mathematical life course". She wrote:

> According to the myth, mathematical talent and creative potential emerge very early in childhood [...] the student proceeds from college to an elite graduate school without a break. In graduate school, the student comes under the tutelage of a powerful mentor, under whose direction he writes a doctoral dissertation that makes a significant contribution to his area of study. As his graduate studies draw to a close, his mentor assists him in landing a postdoctoral research position at a similarly elite doctorate-granting department of mathematics, and afterward he goes on to one or more positions at comparably distinguished universities, where his creative achievements are rewarded with tenure. (p.16)

I was interested to find out whether this 'myth' was actualised in the life stories of the mathematicians I interviewed.

As with any language, mathematics has its own rules of constructions, meanings, communicative devices. That mathematicians in many different countries, and language groups, now share these is not an argument for their 'objectivity' or transcendence but simply a recognition that the applications of mathematics, because of the history of the discipline and the power that it has carried in the past, are now spread globally. In the terms of Joseph

Weizenbaum: "some human actions, e.g. the introduction of computers into some complex human activities, may constitute an irreversible commitment" (1976: 28). Such a 'commitment' exists within mathematics too of course. Even as recently as the 18th century, mathematicians in different parts of Europe were devising different systems with different symbolisations that would, eventually, be resolved into one. Such a process, far from undermining the socio-cultural derivations of the subject, emphasises the inability to separate the mathematics from those who are creating and using it. If anything, it reinforces them, pointing to the pervasive culture of the discipline itself as a powerful factor in how it functions, how meaning is made, recognised and valued, and who achieves influence. And, the 'myth' of globality tends to obscure differences. Writing in 1976, Luke Hodgkin pointed out that

> We conveniently call 'mathematics' one subject today (or we subdivide it into any number of subfields). In fact, though, I'd suggest that we have *two* sciences which go by the name of mathematics, characterized by fundamentally different practices. The distinction is not between pure and applied; I'd rather describe it as between 'old' and 'new' mathematics, where the old includes pure *and* applied as they existed up to the 1930s". (1976: 50, original emphasis)

He went on to

> emphasize the distinctness of these two sciences, which is at a different level from the separation between different branches of 'old' mathematics, for example. It is possible to learn one – in the sense of receiving a full training – and know nothing at all of the other. (Ibid: 52)

Would the mathematicians' practices when doing research support the traditional, transmissive style of teaching so frequently observed, or be rooted in their personal and socio-cultural positioning and a consequent alternative perspective on the discipline? Would I find any mathematicians who, like Reuben Hersh, believed that "mathematics must be understood as a human activity, a social phenomenon, part of human culture, historically evolved, and intelligible only in a social context" (1998: xi)?

From Homogeneity to Heterogeneity

An 'objective' view, on the nature of mathematics, led almost inevitably to a perspective on the discipline as homogeneous. Thus, a second challenge, to homogeneity, derives from the first. Mathematicians are presented in the literature as engaged in searching for The Truth of which, by definition, there can only be one. From time to time, a 'truth' is wobbled off its perch, either by being shown to be only partial, or by being undermined altogether. The blame for this lies, it would appear, not in the mathematics but in the inadequacies of mathematicians. But Brian Rotman pointed out the links between the socio-cultural context and the convincing nature of the mathematics derived by mathematicians:

> Mathematicians believe because they are persuaded to believe; so that what is salient about mathematical assertions is not their supposed truth about some world that precedes them but instead the inconceivability of persuasively creating a world in which they are denied. (2000: 41)

A singular view of mathematics also controls the ways in which it is presented to learners. Both are conceived as homogeneous, the mathematics itself, and the students. The students are often tracked, or setted, in the assumption that reducing the range of achievement makes the group easier to teach because they are more 'like' each other. So, as Jo Boaler (1997a) recorded, conventional lessons are text dependent, questions are broken "into small, atomistic parts" and students are guided "through any mathematical decision-making" (Ibid: 20). Teachers "combined high-definition questions which had one correct answer, with a closed sequencing of content, moving in 'tight, logical steps between one item and the next' (Ibid: 21). Other researchers have found similar patterns in mathematics classrooms. Furthermore, "students held a view that mathematics was all about memorizing a vast number of rules, formulas and equations. They did not believe that mathematics was a rich or varied subject" (Ibid: 36). Alan Schoenfeld (1985, 1988) linked the resultant student behaviour of looking for hints and tips to resolve a difficulty, rather than trying to make mathematical sense, to the pedagogic practices of teachers that emphasise the learning of specific and specified skills or routines. In this way, the mathematics, too, is treated as if it is homogenous.

Abbe Herzig found that university faculty held two beliefs, that mathematics is difficult and that only some people are talented enough to be able to do it.

Implying that many students lack talent, a faculty contributor to her study said: "I don't really think that in most areas to write a PhD thesis you need a really new idea." (2002b: 184) and, in any case, the faculty

> described the qualifying exams as having two main purposes: to serve as a filter, helping students understand if they can cut it or not, and to force them, to absorb and synthesize a large amount of information, certifying that they have breadth of knowledge in mathematics. (Ibid: 189)

That is, even at this level of study, these students were being "force[d] to absorb and synthesize a large amount of information" (Herzig, 2002b:189), knowledge objects ('truths'), that they were expected to acquire and certainly not to challenge. The assumption is that only the highly talented will survive such a regime. The style through which this material was met was of course delivery mode. Again, the assumption was that there was particular mathematics to be learned and that all the students could be treated similarly in that learning. Abbe Herzig reported the PhD students making frequent

> complaints about the teaching of the courses, including the lack of interaction between the instructor and the students, difficulty discerning the important information, incomprehensible lectures, non-English textbooks, and the lack of motivation or connections among mathematical ideas and the mathematical 'big picture'. (Ibid: 190)

It would be unreasonable to understand mathematics as dependent upon those who derive it, located in a particular culture and time, without expecting that different mathematicians would themselves differ in how and what they understood. Such differences across time and cultures are recorded in the literature quoted above although David Bloor pointed out that "The idea that there can be variation in mathematics just as there is variation in social organisation appears to some sociologists to be a monstrous absurdity" (1991: 107). However, he offered examples of

> four types of variation in mathematical thought each of which can be traced back to social causes. They are (i) variation in the broad cognitive style of mathematics, (ii) variation in the framework of associations, relationships, uses, analogies, and the metaphysical implications attributed to mathematics, (iii) variations in the meanings attached to computations and symbolic

> manipulations, (iv) variation in rigour and the type of reasoning which is held to prove a conclusion. (Ibid: 110)

In his Chapter 7, he added a fifth source of variation that he called 'Negotiation in Logical and Mathematical Thought' providing "a richer picture...of the creative or generative powers of thought. A more sophisticated understanding...of just what the compulsion of a logical or mathematical argument amounts to" (Ibid: 131). The example offered to support variation in cognitive style consists of the changing responses over time to the question "Is 'One' a Number?" (Ibid: 110-118).

The view of mathematics as homogeneous is disputed in the mathematics literature as, indeed, some of the quotes above have shown. Nonetheless, a considerable body of research done in mathematics classrooms with teachers and students has demonstrated that the impression given to students is that mathematics is a one right-answer, one right-method, discipline. Would I find heterogeneity or homogeneity amongst the research mathematicians and, in particular, in their attitudes and beliefs about how they practised their discipline?

From Impersonality to Holism

Closely linked to homogeneity, one solution/one method, is the extremely abstract presentation of which so many students, such as those quoted above, complain. Mathematics is triumphant about its abstractness and the removal of the personal from its presentation. It was my expectation, therefore, that an approach that reflected heterogeneity would have to be more holistic and consequently inclusive. Certainly in the mathematics education literature, where researchers report the effects on learning of encouraging students to enquire into the mathematics necessary to resolve their own problems, there is an emphasis on the engagement of the whole person through listening, conjecturing, challenging and constructing argument with others. Maria Bartolini Bussi put it well:

> In order for students to become mathematically literate adults, they must learn to speak 'inquiry math', that is, to ask mathematical questions; to solve mathematical problems that are new for them; to propose conjectures, to listen to mathematical arguments; and to read and challenge popular articles containing mathematical content (1998: 15)

Barbara Jaworkski reported on the results of such an approach in the secondary school classrooms she observed:

> It has been mathematical activities which have formed the basis of meaning-making – negotiation and expression of mathematical ideas, mathematical pattern seeking, raising of mathematical questions, invoking of mathematical images. There was overt emphasis on thinking, which was mathematical thinking. However, I felt that this went beyond the subject to a meta-level of thinking about the act of learning itself. I felt that each of these teachers challenged students to take responsibility for their own learning. (1994: 177)

The notion of taking responsibility for oneself, rather than depending on teacher or text, is linked in the literature to a classroom climate that encourages learner agency within a learning community. An important aspect of agency is the finding and using of 'voice'. Hilary Povey and I linked this to what we called 'authoring':

> ...fostering an epistemological perspective of *author/ity* among teachers and learners will support a renegotiation of the relations of dominance embedded within current conceptions of the nature of mathematical knowledge. Such an epistemological perspective takes the concrete and the personal as the starting point for meaning-making; it recognizes the vitality and significance of dialogue in the process of knowledge construction within a community; it relocates the privileging of the read *and* written in the mathematics classroom into a more coherent approach drawing upon speaking, listening, reading and writing and emphasizing meaning construction and negotiation; and it nurtures within that community an ethic of care and of personal accountability. (1999: 243/244,original emphasis)

This is a perspective on authorship that begins with the informal and only moves to the formal as a record of readiness to share publicly. It is particularly well suited to a community of learners. But researchers are also learners, and they are also under an obligation to publish. Again, as queried earlier, I was curious to find if there would be affinity between this stance in mathematics education, and the stance of the mathematicians in the study.

An important aspect of community is the rejection of both isolation and emphasis on the individual. Before I began the study, I did not know if the public image of the mathematician working alone would be confirmed. This image I called the 'Andrew Wiles' effect after the mathematician who over ten years of sustained work, solved Fermat's Last Theorem. The story is told in Singh (1997) where the impression is given that much of this work was done in isolation.

I had no reason to expect that the mathematicians would not confirm the image. However, in school classrooms, mathematics educators advocate the kind of community described above as being more conducive to learning and this is reinforced within both the mathematics education and the general educational literature. However, it is not the kind of environment that is frequently found in schools or in universities although I know of some departments in both the UK and the USA, and in Sweden, where experiments of this kind are ongoing. Would this constitute a major difference between learners of mathematics and researchers?

The literature on intuition and aesthetics, and their role in the creation and use of mathematics, then feeds directly into a need to recognise, incorporate and make substantial the whole person of the mathematician into their mathematical production. But, additionally, the recent works of Brian Rotman (2000) and Robert Thomas (2000, 2002, 2004) are interesting from this perspective. In drawing connections between narrative and mathematical proof, Robert Thomas' work can, in my view, be read in two inter-connected ways. One way, adopted by Robert Thomas himself, is to examine mathematical argument as a narrative, which he does very convincingly (See Thomas, 2000, 2004). But connected very closely to that is the role of the mathematician as narrator. David Bloor pointed out that

> The higher interpretive principles which embody meaning, proof and rigour correspond to the explanatory theories, paradigms, research programmes and metaphysical frameworks of the natural scientist. There appears to be no reason why mathematics should be treated any differently to the empirical sciences. (1991: 129)

And Brian Rotman warned that:

> once the Person is acknowledged as vital to the mathematical activities of making and proving assertions, it becomes impossible to see [...] a domain of mere psychological/motivational affect, ... be[ing] jettisoned as soon as the real, proper, rigorous mathematics ... has been formulated. (2000: 54)

How much, I wondered, would the mathematicians I interviewed relate to the kinds of arguments that reject isolation and, its consequence, competition, and embrace membership in a community of practice with all the implications for behaviour, attitudes and values? Would the mathematicians speak of themselves holistically, feelings, cognitions and community membership all interacting, or see themselves only as individuals functioning as isolated units?

From incoherence to the coherence of connections

Learners experience mathematics as an extremely fragmented discipline. It is fragmented internally in that the presentation in texts, and many courses, usually separates one item of content from another and not all teachers ask students in what ways aspects of mathematics are connected or relate to each other. Disciplinarily, it is also fragmented, certainly in Anglophone countries where departments of pure mathematics, of statistics and of applied mathematics are often unconnected within the same university. And yet, when mathematicians write about mathematics and even in public documents (for example the Non-Statutory Guidance to the UK National Curriculum, early version), reference is frequently made to 'networks of knowledge'. Mathematics is sometimes compared to a map but learners have the greatest difficulty in working out the key to reading or using it.

It was my belief that fragmentation was counter-productive for learners and that its sole apparent utility was in facilitating the use by teachers of exercise lists for practising mathematics and the asking of disconnected, fairly meaningless, questions on examinations. When syllabi, one version of mathematical stories, are put together in disconnected fragments there appears to be an un-stated expectation that the learner themselves will achieve a coherence which, if it exists at all in the syllabus, does so only very covertly.

The drive to search for and establish what I came to call connectivities contradicts the need-to-know assumption that underlies most mathematical syllabi and is a rationale for the delivery mode of transmission teaching. And thinking of the stories or narratives that fail to be told in textual or oral deliveries in classrooms makes even clearer why so many learners of such fragmented syllabi fail to encounter mathematics. The incoherence of the narratives, in part, is a function of the reification of mathematical objects but:

> mathematics is not about mathematical objects but about the relations that hold among them and which they, like fictional characters, are invented to support, to allow us to speak in a way that is interesting because it says the things we are interested in saying. (Thomas, 2002: 4)

The expectation of the importance of connectivities arose, for me, from widespread references to coherence and connections in the mathematics literature as well as from the educational and feminist literature that stresses the need to learn in a connected way. I was, therefore, interested to discover what the mathematicians I would interview might say about connections in mathematics and about mathematical coherence. If, similarly to those mathematicians whose work I had read, the participants in the study emphasised the importance of connectivities, I would need to look for new explanations as to why those who prepare and use syllabi for the teaching of mathematics continue to support the current fragmentation.

MAKING MEANING – COMING TO KNOW

Answering these four challenges by an emphasis on socio-cultural derivation, heterogeneity, on the whole person *in* the mathematics and within the community and on mathematical coherence, led me to a perspective on mathematics that is person- and socio-culturally, rather than discipline, driven, a mathematics of coming to know through meaning-making.

With an emphasis on meaning-making, the model, as derived, had its own coherence and connections and, I believed, explanatory power. If the mathematicians whom I interviewed supported, and were supported by, the categories of the model, it seemed to me that there would be a substantial argument for approaching the learning and teaching of mathematics as a meaning-making, rather than meaning-transferral, enterprise. It was my hope that a gap, between how mathematicians themselves came to know and how they promoted learning in others, if confirmed in the study, would help to explain student disaffection with the discipline, their difficulties in learning it, and the gender bias in those who take it up as a career. Identifying such a gap is, I believe, the first step in bridging it. The study was set up to explore whether or not such a gap existed and, if so, what were the features of the mathematicians' coming to know processes that might be applicable to anyone learning mathematics.

The empirical study

To offer grounds for trusting the outcomes of the study, I now describe how the empirical study, to test the model against the descriptions of mathematicians as to how they come to know mathematics, was designed, executed and the data were analysed. I undertook the study in 1997, with thirty-five female and thirty-five male participants who were in career positions as mathematicians in twenty-two universities in England, Scotland, Northern Ireland and the Republic of Ireland. I wanted to explore how they came to know mathematics, their feelings and career experiences that might have influenced how they thought about knowing mathematics. Jacques Hadamard (1945) had pursued some similar questions. However, that work was based on anecdote and questionnaire (as was the famous Poincaré 1913 lecture). I wanted to use the flexibility of the personal conversation to try, systematically, to unpick the attitudes, beliefs and practices of the research mathematicians when they talked to me about their engagement with a research problem from inception to publication. It was my intention to compare what was said with the model I had devised and which has already been described.

Seeking the participants

I knew from previous work that female mathematicians were more difficult to find than males so I first approached female mathematicians using personal and institutional contact ('snowballing') and electronic networks. I invited them to be interviewed about their 'life history' as mathematicians, especially their feelings about knowing mathematics and how they come to know. Once a female mathematician agreed to be interviewed, I invited her to find a male 'pair' who was agreeable to join the study, preferably in her institution (for ease of my travel/time). Although the choice of male, and the reasons for choice, were left to the female participant (and conveyed to me), most chose someone at a similar level of the hierarchy to themselves and/or in the same mathematical area. Three of my pairs were marital partners, usually as a result of the female having difficulty in finding and persuading a colleague to participate! This might mean that the males in the study are less representative of male mathematicians than the females but, if so, I did not get the impression from any of the participants that this affected their perspectives on coming to know mathematics. In a very few cases, I was directed towards particular mathematicians as having something interesting to say about coming to know mathematics and when they agreed to participate in the study, they were asked to find a 'pair' of the other sex. The seventy participants were well spread in terms of their employment status (Table 2.1):

Table 2.1. Employment status of participants

	Postdoc	Lecturer	Senior Lecturer	Reader	Professor	Senior Research Officer	Research Fellow
Females	1	19	7	3	3	1	1
Males	1	17	9	2	6		
Total	2	36	16	5	9	1	1

One of the two female research-only participants, together with some of the lecturers, female and male, and both post-doctoral fellows, were not in permanent posts. Although there were twice as many male as female professors, the numbers are very small and not necessarily significant although female professors of mathematics in the UK, at the time of writing, constitute only 2% of the total (nine individuals). The Oxbridge system is such that anyone who is not a Reader or a Professor is listed as a Lecturer including some who have senior administrative responsibility and others who are Tutorial Fellows.

Design

I met all except six participants face-to-face in a place of their choosing, usually their office. The six remaining were interviewed by telephone. The shortest interview was one hour, the longest two and a half hours, and most were between one and a quarter and one and a half hours. The telephone interviews were among the shortest. I attribute this to the lack of personal, non-verbal information for both parties to the conversation. The interviews with the female participants tended to be longer than those with the males, which might have been due to the fact that I am a female. Prior to the interview, all participants received a list of the areas that I wanted to explore about them and their coming to know mathematics (see Appendix A) and a brief questionnaire to provide factual information (see Appendix B). I invited them to delete anything that they did not wish to discuss. Nobody requested any deletion. The list served, during the interview, as a checklist that the various aspects had been addressed and that the interviews covered similar grounds although it was not used in any sense as an agenda. I agree with Simon Goodchild, in rejecting any suggestion that an interview-based approach is unsystematic. He said:

> the natural flow of the conversation was an important aspect of the research design. It was necessary to make the encounters ...as smooth as possible talking to them within their own context, seeking to enter their setting with minimal disturbance. These are essential features of research aspiring to ecological validity. (2001: 104)

I also gave the mathematicians details about the conduct of the study, including my commitment to return to them notes of the interview for agreement/amendment or deletion prior to analysis. They were, of course, assured of confidentiality. All the interviews were tape-recorded and, additionally, I took notes. The interviews followed a 'life-history' format and were conducted in conversational style. If some aspect had not been addressed, I would raise it by saying something like: "I notice that you haven't mentioned ...Why is that?" Otherwise, the direction and content of the interview were guided by the participant and not by me.

Data analysis

The interviews were transcribed in a shortened form (that is my voice was removed) and, as agreed, returned to the mathematicians for their comments. The data were entered into NUD*IST (Non-numeric Unstructured Data-Indexing, Sorting and Theorising: QSR 1994) which is a software package for the analysis of qualitative data. Quantitative information was recorded on a spreadsheet using EXCEL. The categories applied to the analysis, in the first instance, were those of the model but of course many more categories were derived from the data. Persistent listening to the tapes, and staring at the transcripts, ensured that I was immersed in the data in line with Cohen and Manion's (1989) advice to collect:

> as much data as possible relevant to a particular topic regardless of where it occurs in the interview material. The method is exacting because it requires very detailed knowledge of content and may entail going through taped interviews several times. (p249)

The final classificatory system contained nine categories which were: History, Intuition, Thinking Styles, Research Style, Mathematics is, Publishing, Feelings, Mathematical Problems and Culture. Each one of these was broken down into sub-categories of which History was probably the most complex containing fourteen such, some of which, themselves, were divided into further sub-sub-categories. For example, the sub-category Community, was a response to a question such as: *With which community would you identify, for example to what conferences do you go, for which group might you look at the bar?* I separated the responses into Applied, Pure, Statistics, Operations Research and Applied Probability. These were then sub-categorised into the various specialities that the mathematicians mentioned (see Appendix D for the full tree diagram).

Although some aspects of the data are tied to the UK and Ireland, anecdotal evidence from other countries suggests that in many other respects the situation is similar from country to country. Both the model and the resultant data from the study presented in this book have been the basis for seminars given in many parts of the world and nowhere has a single mathematician queried the match between their own experiences and the experiences I am recounting together with my analysis and interpretation. Indeed, on the contrary, some mathematicians have expressed surprise that research was necessary to reveal results that are so 'obvious' (that is familiar to them and within the community of mathematicians).

SUMMARY

My focus was *not* a knowledge-based enquiry. I was not concerned with the knowledge that the mathematicians were engaged upon building although I received many extremely interesting tutorials. I was interested in *how* they came to know and what influences different strategies of knowledge building might have on the achieved knowing. The epistemological model that I set out to validate proved to be remarkably robust when matched with what the seventy research mathematicians said about how they come to know mathematics.

However, in analysing the very large database, it became clear to me that I had to distinguish between the mathematicians' beliefs, that is *what* they think mathematics is, their practices, that is *how* they approach their research, and the seductive nature of their social positioning into what *everybody knows*. Finally, although I had not set out to discuss teaching and learning with them, inevitably there were aspects of our conversations which provided data on how they understood their roles as teachers and as learners and the data contain a lot of material which can be interpreted (with all the caveats necessary to that) to throw light on the links between epistemology and pedagogy.

In the next chapter, I explore the personal histories and beliefs, of the mathematicians, using and organising the information I collected from them in order to answer the question: Who were they? From their histories, I gathered details about their backgrounds. To what extent were the experiences of the mathematicians consistent with the 'developmental myth' that mathematicians are born, not made, or, like the women interviewed by Margaret Murray (2000), did they also deviate from it? They told me about their families, of the schools and universities they had attended, whether or not they had role models and, if so, who these were and what influence they exerted on their choice of career. I sought to find out if there were any female/male indicators about these influences, the point at which they opted for mathematics, and their choices of area within the discipline. All these and many more questions are answered in the data.

3

Who were these mathematicians?

> I drew arrows in different colours to follow the path of her lover's mathematics, and hers. I used single lines for her, and double lines for his. I used all the colours of the spectrum. Slowly the sad empty room became as bright as a rainbow. It was like drawing two different journeys across a map, her journey and his. It took the rest of my adolescence. And though I joined up many arguments, by the time I'd left school, I could see how different two journeys could be. (Woolfe, 1996: 282)

In the past, many publications have dealt with the question "What is mathematics?" Indeed, Reuben Hersh (1998) produced a delightful book that he called *What is Mathematics, Really?* However, there have been few investigations into "Who is a mathematician?" other than someone who does mathematics. (In her book Women Becoming Mathematicians, Margaret Murray (2001: Chapter 7) responds to the question: What does it mean to be a mathematician?) I hope that this chapter will help to address the lack and provide some surprises along the way. In particular, I challenge the stereotype of the mathematician who shows aptitude from a very early age and who comes from a mathematically 'connected' background, neither of which is supported in the literature nor in my study.

At the start of the interview, I asked each participant to describe their 'historical trajectory' in becoming a mathematician (see Appendix A). The data on which this chapter is based, in part, came from what I was told. The participants also completed factual questionnaires (see Appendix B). In this chapter, I discuss those aspects of the histories that struck me as being surprising or unexpected and consequently noteworthy of report. These related to what influenced the original choice to become a mathematician and when, their social class and educational backgrounds, their career paths. Since their beliefs about mathematics were likely to be closely intertwined with who they are and their consequent practices, I also present the range of beliefs that they held on what is mathematics.

HISTORY

The Myth of the Mathematical Life Course

As I mentioned in Chapter 2, there is a myth associated with mathematicians that they are born not made and that their life histories are necessarily particular to mathematics (Murray, 2001: 15-19). Interest in mathematics, and demonstrated talent at it, is said to emerge very early and to overwhelm other interests to such a degree that, although mathematicians are perceived as intense about their discipline, their range of interests is considered to be narrow. (The only counter-indicator to this is the widespread observation that interest in mathematics often goes with interest in music, although not necessarily in the other direction!) Margaret Murray proposed that this is both a self-fulfilling prophecy *and* a post hoc justification for why some people remain and others leave mathematics.

> The myth has been, and continues to be, aggressively perpetrated, particularly in interviews, autobiographies, biographies, memoirs, and reminiscences of mathematicians whose careers reached their full flowering during the postwar expansion. (Murray, 2001: 18)

Indeed, the myth of born not made is alive and well in school and university classrooms, believed and perpetuated by teachers and taught. How did the mathematicians I interviewed respond to the question of when they first became interested in mathematics?

There is some possible lack of clarity in the data, remembering that they were speaking to me about their early history[1]. There were those who made a distinction *between wanting to study mathematics and wanting to become a mathematician.* One mathematician explained this difference:

> *I didn't know that it was possible to be a mathematician until I went to university to be interviewed and I realised that the people on the other side of the table were in a career as mathematicians. I knew that I always wanted to do mathematics but didn't realise that I could be a mathematician until I met some.*

Another possible lack of clarity lies in the blurring of school success, and achievement at mathematics, with 'talent': *I was already good at mathematics at school. It was the subject that required the least effort but then, to be honest, most subjects didn't require a lot of effort.* Perhaps for this mathematician, and a number of others, school was not challenging but this does not explain the

[1] Throughout the book, direct quotes are shown in italics.

move towards mathematics. Many of the mathematicians acknowledged the role of others who encouraged, nurtured or even 'turned them on' to mathematics. Often, this was a teacher (see below).

It is impossible to say, therefore, how much of their career choice can be laid down to 'talent' and how much to environmental influences. Nonetheless, of the seventy mathematicians, only 18 spoke about early influences: *I cannot remember any time when I didn't want to do it;* but 26 came to a realisation during the secondary school years: *I think in fifth year a lot of other people found it quite difficult and I got switched on to it.* For some, it was as late as when they chose to specialise at Advanced level (16-18 years); and for 5 their choice was only confirmed at university: *By the end of first year, I realised that I wasn't quite bottom of the heap. By then, it was becoming clear that if I was going to become an academic, then it would be mathematics.* These data, therefore, do not substantiate a reading that mathematicians are born, at least in the sense that they are uniquely drawn to mathematics from a very early age. One said: *I certainly didn't wake up one morning and think I want to be a mathematician.* And another: *I certainly cannot say that I ever set out to be a mathematician. It was more a drifting into mathematics.* But there were others who claimed: *I would have had to make a decision* not *to do mathematics.*

Because they contradicted the stereotype, I was particularly interested in those who described alternative influences, including the one mathematician, who acknowledged the role of the UK Royal Institution Christmas Lectures in causing a switch from science to mathematics: *I was in my first year of the Sixth Form, 16 or 17, and it was the Royal Society* [sic] *Christmas Lectures given by Christopher Zeeman. I decided I wanted to be a mathematician and that I wanted to study at Warwick.* Some mathematicians highlighted the close links between physics and mathematics in claiming that their first choice had been physics, or science more generally, and that they only subsequently chose mathematics. This was the case with the 2 males who made a switch from physics to mathematics during their university studies. One replied to the question of when he had decided that mathematics would be his chosen career subject by saying:

> *At the age of 25 or 26 when I started my Ph.D. Before that it was not my idea to become a mathematician. I wanted to be a theoretical physicist or something like that. I thought of mathematics as a tool used by the physical sciences or as a language, possibly coming from utilitarian ideas. I thought mathematics was rather easy and not challenging. Physics was more challenging possibly because I could make more stories out of it. Since then, I have changed my mind and think that mathematics is more amenable to have constructions and ideas and create stories probably more than physics. It wasn't a sudden change. It came slowly.*

The 3 females who also made the decision to become mathematicians at university had chosen to do mathematics degrees but, apparently, this choice had not committed them to a career in mathematics. One explained:

> *In my third year of university I decided to major in mathematics but I didn't really think of myself as a mathematician. Then I did my honours year and still didn't really think of a career in mathematics. It was only when, by a bit of luck, I got my first job as a mathematics research assistant and discovered that I liked researching maths and could do it that I thought of making a longer term job of it. So that is when I would consider myself to have become a mathematician. I think I didn't really realise there were mathematicians. I just thought that scientists would do some mathematics just as part of their work. It wasn't until I did that job that I thought of doing a Ph.D. because it gave me the confidence to think I could not only study maths but also do research. That's when I thought I would like to do a Ph.D.*

Thank a Teacher

Almost all of the mathematicians mentioned that they had been good at mathematics at school. One exception said *I was rotten at arithmetic at school and I had no particular liking for it. I wasn't one of these ones who wanted to do it since primary school.* Even if they found mathematics easy, that did not necessarily mean that they originally wanted to pursue the subject, or even, at school, that they knew that a career in mathematics was possible, as one mathematician earlier explained. And being good at mathematics did not necessarily mean that they were good at, or enjoyed, all aspects of the subject: *I remember taking to classical Euclidean geometry like a duck to water but really struggling with algebra.* Nonetheless, what I found interesting was the number of mathematicians who remarked upon the role of a teacher in influencing their attitudes to, or experiences of, mathematics at school. 32 mathematicians specifically commented on the effect on them of a particular teacher; 17 of these (just over 53% of the women) were female. Speaking of experience of mathematics in secondary school, one mathematician said *I owe a lot to my maths teacher who was tremendous. I had her all through the first five years and for pure maths in the sixth form.* And another spoke almost identically: *I was lucky enough to have the same mathematics teacher for nearly my whole secondary school period. He had a very strong influence on developing the mathematics.* Murray (2001) showed that this is not particular to my study:

> Nearly all of the women interviewed mention that they had at least one teacher at the college level who sparked their interest in mathematics and made an investment in their present and future success. In some cases, they

attribute their determination to go on in mathematics to the influence of just one special teacher. (p.99)

She described (Ibid: 72-99) high school, for the women mathematicians she interviewed, in terms of: an end to isolation, a place of intellectual adventure, a place of limitations, a place of encounter with mathematics and of gender barriers: "From elementary school on through high school, college, and beyond, girls and women have often been told that they cannot or should not do mathematics and science" (Murray, 2001: 92)

It is in the sixth form, at age 16+, that mathematics in the UK becomes optional for the Advanced ('A') level examination. Another mathematician remarked:

I wasn't particularly interested in 'A' level mathematics although I did have a maths teacher who was forever getting in trouble, and he did this bizarre thing of letting us talk to each other and he used to leave us to it and let us get on with it.

A female mathematician organised her preferences for herself:

I went off to grammar school and did modern maths for five years. In the sixth form, everyone did traditional maths and I thought I didn't want to do that, I thought it was all applied and not as much fun, so I went to the local boys' grammar school and I asked to go there and do modern maths. So they adopted me and I did maths and further maths in the boys' school and physics back in my own school. It was great because I was an individual and a bit of an oddball and in the further maths there were three to start with and then it got down to two and it was only two because it was improper for me to be on my own with a male teacher. So I was being treated as an individual and that was great.

Another female mathematician summed up, for me what many said about the impact of a teacher; but she also explained how that impact was realised, for her, in a way of thinking that remained with her:

When I was young, I had no interest at all in mathematics. Long multiplication of pounds, shillings and pence we had to do ad nauseam. It was like witchcraft - one had to know the spell - I had no interest whatsoever until I was about twelve. Then, for the first time I believe, I was taught by someone who was qualified in maths. She probably qualified in about 1920 when there couldn't have been many women qualified in mathematics. Possibly I would never have become interested in mathematics except for her. She explained why things happened. The previous teachers had just taught you the rules. I had always loved history and English but mathematics had been totally stultifying. We started

> *learning Euclidean geometry which I absolutely loved. I liked it so much that when I used to go and stay with my Irish grandmother in the Summer, I can remember going to sleep night after night with geometrical figures going around in my head. And I was trying to see what was going on. I was visualising. So, from no interest at all, suddenly mathematics became the thing I was most interested in. The thing that came as a complete revelation was that you could work everything out – the underlying reasons – just by thinking about it. You needed to know very little. By degrees, I worked out why things made sense. That has never left me.*

Many of the issues raised by this mathematician address the kinds of conditions in mathematics classrooms about which students complain as well as aspects of the learning of mathematics that are important to, but rarely found, in student experience. I will pick these issues up again in Chapter 10.

Home Background

Of the 70 mathematicians, 20, 9 females and 11 males (nearly 30%), came from a working class family where there was no history of university attendance: *I come from a working class background and while my parents encouraged me, and were interested in education, they couldn't help me. There was no academic history. It crept up on me with a bit of a surprise.* Four more were middle class but with no university connections. A further 31 mathematicians, 17 females and 14 males, had parents in middle-class occupations. Of these, 27, 13 females and all 14 males, mentioned at least one parent who had been university educated; 4 sets of parents (i.e. both mothers and fathers) had undertaken (some) studies in mathematics at university of which 2 sets were in academic roles as mathematicians. In both of these latter 2 cases, the participant mathematicians were female. On the remainder, I had no background information. As can be seen in Table 3.1, fathers were mentioned much more than mothers. This does not mean, of course, that mothers did not have a degree. However, it does mean that mothers were not reported as being as influential as fathers and one (male) mathematician went so far as to say: *My father is an actuary and he went out and taught himself maths. My mother has a maths degree but she was never quite the mathematician of the family.*

Table 3.1: Family connections with university education

	No family uni. Connections	Both parents uni. Educated (not maths/ engineering)	Father maths or physics degree	Mother maths degree	Father engineering or other related degree	Mother other uni. Degree	No uni. Connections (middle class)
Female	9		7	3	3		4
Male	11	3	2	1	6	2	
Total	20	3	9	4	9	2	4

On the other hand, another (female) mathematician said:

> *My mother also encouraged me very much. She was ambitious. All the pushing from home came from my mother who, in any case, was a lone parent throughout the war years. I was six and a half when my father came back from the war. My mother supervised homework. Mum tried to teach me to solve problems with algebra in the steam on the bathroom wall.*

In the most recent figures, and consistent over time, 25% of students entering UK universities are from the working class. Nonetheless, I was surprised that the number in my study from a working class family (28.6%) was so high. My surprise caused me to reflect on why this ratio is so robust, especially with new entrants, given the prominence of governmental strategies to improve the situation. As already indicated, many of the mathematicians commented on how grateful they were to a particular teacher and such remarks were often made by those who came from backgrounds where there would be minimal knowledge about the university sector and how to support or inform a future university student. At one time, identifying and nurturing a working-class child through to university entrance counted for a lot in schools and doing that in a high-prestige subject like mathematics would have been additionally important. This is probably still the case but two things have changed. One is the climate in schools and the other is the availability of mathematics teachers. Although there are still exceptional teachers, the number has dwindled under the pressures experienced in schools and the diminishing number of those qualified and trained to teach mathematics (see Johnston-Wilder, 2000 which addresses the "current problem with teacher recruitment and retention".) And there is a crisis, internationally, in attracting students to study mathematics and, consequently, to go into teaching. The Economist reported on June 30, 2001 that "Last year, some 52 universities failed to meet their targets for student recruitment" (p.34) and Simon Singh in The Daily Telegraph (September 26, 2001) pointed to the specific problem with mathematics in an article entitled "Is the number up for maths?" (And see Chapter 1.)

But the role of university mathematicians in influencing who remains in mathematics cannot be ignored and, too often, students complain about the quality of teaching and, indeed, even of PhD supervision (see, for example, the work of Abbe Herzig quoted in Chapter 2).

> Mathematicians in universities have great power to influence what is learned and how it is learned in the discipline. They do this, in part, through the respect which society accords their discipline, and consequently those within it…But these mathematicians are also responsible for teaching the next generation of teachers and, consequently, for contributing to the

definition and history of the practices which are seen to be appropriate to the communities that they touch. (Burton, 2001b: 589-590)

Influence of Place in Family and Role Models?

In the gender literature, a considerable amount has been made of whether or not the female mathematician is an only child, or the eldest in the family and on the influence, particularly on females, of a role model. I was interested, therefore, to know whether or not these were features for my participants. Table 3.2 gives the answers to the question: Are you (a) an only child? (b) an eldest child? (c) placed in the family how? Please use B (brother), S (sister) and X for you and show oldest to youngest in line, e.g. BBXS.

Table 3.2: Position in family

	Only child	First in family	Not first in family
Female	1	19	15
Male	4	16	15
Total	5	35	30

From these numbers it appears that half of my participants, whether they were female or male, were first in the family but that only children do not feature significantly and certainly not for the females.

The question: Was anybody in your family connected with mathematics in any way? was answered affirmatively by 20 mathematicians plus one who spoke of her father, and other paternal relatives as having *been good at it.* The connections cited, in addition to those already mentioned, were grandfathers and fathers who were mathematics or science teachers, mothers and fathers who had done some studying of mathematics at university and parents who had a high respect for the discipline of mathematics and, consequently, encouraged their children in that direction. 10 females and 4 males cited engineering as a mathematical connection, through fathers, grandfathers or brothers. But, although for most these connections were positive, this was not the case for all as one female mathematician explained:

> *My father was a mathematician and I hadn't really wanted to do maths. He used to tell me frequently that I was the most stupid person he had ever met at maths. He told all of us that and we all did maths. He found us very frustrating to explain to.*

Career Path

In recounting the passage from undergraduate, through postgraduate studies to career, the mathematicians made me aware of a gender difference between those whose path was smooth and continuous, and those for whom it certainly was not. 20 of the mathematicians gave details of smooth passage, without any breaks, although not necessarily directly into permanent posts. Of these, 7 were female, 13 were male. Below are the kinds of comments that were made, first by a female, then by a male:

> *As a post-doc., I went to America which wasn't an academic institution, it was a research institution and that was fantastic...it was a limited time appointment and people moved mostly to academic appointments. I was quite ready to become a lecturer, it was a natural step and I didn't want to do it in America, so I came here.*

> *Shortly before I got my PhD I got a research fellowship at Oxbridge. I stayed for one year and then went to America as an Assistant Professor. I was there for seven years and then got a teaching fellowship at Oxbridge which became tenured.*

By comparison, 10 of the women described career breaks. In some cases, these were associated with completion of formal education: *I actually went to university 8 years after I left school.* No males reported this. Sometimes, the cause given was marriage and/or children. *At the end of my undergraduate career, I married and started a family. I had seven years off with the children and then came back to do a PhD.* Such differences have been extensively reported in the general literature (for example, see Acker & Feuerverger, 1996, Heward, 1996, Morley & Walsh, 1996) as well as in the mathematical literature (Murray, 2001). A not untypical case, was a woman who had dropped out of Oxbridge, had children and spent 10 years outside of mathematics while her husband completed his qualifications and began an academic career. She reclaimed her academic interests and confidence through doing voluntary work and gaining a language 'Ordinary' level GCE, after which she successfully took a mathematics degree and then obtained a PhD. Another female reported that she was told that she was doing a PhD *for the love of the subject not because there were any career opportunities.* One male spoke of taking a career break, in his case to travel and decide on his future, a purpose that is encouraged and valued.

Temporary appointments, often post-doctoral fellowships, are the first step for anyone hoping to enter a career in academe after completion of the PhD. Both females and males went along this career path although fewer of the males commented on the stress of being temporary compared with the females, one of whom said: *Then I went to the States, visiting teaching positions at Chicago and then Columbus Ohio, but I was worried about what would happen next.*

In speaking of their careers, the influence of Oxbridge (I refer to Oxbridge, the Universities of Oxford and Cambridge, in order to protect my participants' anonymity) was conveyed very vividly. Thirty of the participants, almost half, had experience of Oxbridge as career mathematicians and/or as students.

Table 3.3: Experience of Oxbridge

	Oxbridge only undergraduate	Oxbridge only postgraduate	Oxbridge undergraduate and postgraduate	Graduated elsewhere, working Oxbridge
Female	6	1	5	1
Male	2	5	5	5
Total	8	6	10	6

That this is not an artefact of this particular study is borne out by Halsey (1992).

Fifteen mathematicians, 7 men and 8 women, expressed feelings about their experiences at Oxbridge. Of these, only 6 were either neutral or positive (4 males, 2 females). Those, 6 women and 3 men, making negative comments cited snobbery, certain students being 'chosen' (i.e. favoured), bad teaching, put-downs, lack of positive indications as to progress and possibilities and being 'overawed'. Typical was the comment: *My experience of Oxbridge was that it was very hostile and dismissive of anybody who didn't have a first degree from Oxbridge and that didn't help.* In particular, comments were made on the 'sink or swim' feelings that many female participants had and the lack of support or nurturing when they were undertaking their studies, whether these were undergraduate or postgraduate. Typical of these was:

> *At Oxbridge I felt that I was being put down all the time and I am still quite angry about it. I could have done so much more with a bit of encouragement.*

Another female told the following story:

> *I went to Oxbridge but I left after a year and took a year out. I had found that year really hard. I thought the mathematics was beyond me. I think their system is very bad for a certain kind of person. They don't do any nurturing. Your main source of knowledge was the hell of one-to-one tutorials. I used to feel ill every week before seeing my tutor.*

A year later, at another university, she *started really enjoying mathematics. Basic things that I hadn't been able to grasp at Oxbridge I got interested in there.* She completed a Master's degree and then a PhD and, when I interviewed her, was in a permanent university position. In telling this story to a number of mathematicians at Oxbridge, I found there were two 'takes': I offered the interpretation that this mathematician had justified the confidence Oxbridge had shown in her originally by giving her a place; however, I found that an alternative interpretation was that she couldn't 'hack' Oxbridge, that is she was not up to it, and 'needed' a more benign, less challenging department. She was certainly very explicit about the aspects of Oxbridge that caused her to leave (as, indeed, were many others who recounted negative experiences) but that seemed to me to be more a commentary on the cultural climate of the Oxbridge departments than on her 'quality' as a mathematician. (The negative aspects have been mentioned above but are dealt with in greater detail in Chapter 10.) Negative experiences such as these were confirmed in a recent report of a study conducted at Cambridge where the mathematics department was found to be particularly at fault (see The Guardian, 2003).

Unfortunately, recent studies with Advanced level students of mathematics and graduating students in a prestigious mathematics department confirm that sentiments such as those expressed above predominantly by female mathematicians continue to be expressed by students, particularly female students (see Chapter 2). And I know that while the emotions are just as implicated in the mathematical thinking of graduating mathematics students, for the majority the feelings are not pleasure, but pain. In another study:

> One graduate of a most prestigious university said:
>
> Never having the chance to really understand things fully. I always felt like I was sinking slowly into a pit of non-comprehension. (Burton, 2002: 165)

When asked the question of what were their (mathematical) plans, now that they had their degree in mathematics, 54% of the female graduates and 25% of the males responded along the following lines, first a female, then a male:

> *I intend to have nothing further to do with mathematics. I have found my experience here very demoralising and think that although I may have gone on to further study had I attended a different university, I have no such intentions now.*
>
> *I intend to avoid mathematics as completely and thoroughly as I can. I feel that my undergraduate experiences have been valuable in giving me a grasp of logical, thorough, ways of dealing with problems, but have also instilled*

> *in me a deep and lasting hatred of pure mathematics. 3 years spent in such a nitpicking subject has been enough to drive many people I know insane.*

Feelings such as these have serious implications not only for the young people expressing them, but also for the many others who share them, including the mathematicians in this study. Further, the implications are carried into the teaching profession *and* the discipline; the costs to those concerned and to the country of such negativity are explored in Chapter 10.

In summary, then, I found no evidence to suggest that mathematicians were born, not made, many of the mathematicians, particularly those from non-academic families, acknowledged the influences of teachers on their choice to enter the discipline, place in family and family connections did not appear to be as significant for females as the literature has implied but the career paths of female mathematicians did deviate somewhat from those of males and Oxbridge played a very important role in many of their careers. I return to some of these points in later chapters.

BELIEFS ABOUT MATHEMATICS

One of my interests was in similarities and differences between female and male mathematicians. I began the study with a conjecture that the socio-cultural system that produces mathematicians is far stronger than differences in gender. I therefore expected to find a high degree of convergence, about the nature of the discipline, in the beliefs of the female and male mathematicians I interviewed. In fact, what I found was both heterogeneity *and* homogeneity. Great heterogeneity was disclosed in the many different beliefs about the mathematics that they come to know, including some, a minority of about 10%, who did not adopt an objectivist, positivist stance with respect to their epistemology. Gender did not seem to feature as relevant to these differences. It is, however, important to remember that, while talking about their beliefs, these beliefs were attached to their researching practices, not to the discipline as a whole. As an example, two mathematicians said:

> *I think it is very common for people not to question mathematics. What is there is true absolutely. When you get further in, we find that everybody is standing on the top of everybody else's shoulders, and there is no uncertainty. Then the bombshell falls. It doesn't now matter to me that it isn't certain. It is the challenge of building the tower not whether or not the tower will fall. You have theorems which you can neither prove nor disprove.*

Complex numbers are just out there. You couldn't have a good understanding of anything in mathematics or physics without using complex numbers.

At the same time, whether espousing Platonism, formalism or neither, there was a uniform philosophical commitment to an absolutist view of mathematical *knowledge* as fixed, certain and a-personal, a singular view of mathematics as if it were without the 's'. Nonetheless, within this philosophical homogeneity, the mathematicians to whom I spoke adopted a breadth and diversity of positions. This diversity is represented in the quotes below. For example, some female participants said:

Some of the time I am a Platonist. As mathematicians we want to believe that it is objective but I guess rationally I think it is socio-cultural and emotionally I want to believe it is objective.

I think mathematics is the product of people. It does turn out to be extremely useful. But even utility is a cultural product.

There is something too strong in it for mathematics to be considered as a cultural artefact.

A male expanded on this same theme but in a way that demonstrated that, for him, there are multiple possibilities:

If somebody said to me is mathematics objective or subjective, I would say subjective. If I were asked if mathematics were Platonist or formalist, I would say formalist but there is a duality because the way we deal with mathematics is more Platonist than formalist. But the issue of fashion and fad and reputation by association, prestige - there is a game which is being played and you make it hard for yourself by not playing it. I do look at scientific facts as complete garbage but I do want it to be a 'truth' - I like the way it is. There is a desire to be a Platonist, to accept these ideas as 'truths' but there is a contradiction that I cannot really do that. On the one hand, I agree with the cultural dependency argument but, on the other hand, I do find appealing the idea that it might be possible to build something which would be meaningful to some other race of people or creatures.

And another presented an argument consistent with, and familiar from, the mathematics literature:

I do see mathematics as a coherent thing...I don't think mathematicians can just make up a set of axioms which will lead to something new. I see grand traditions, mathematical physics, number theory, geometry, things coming down

> *to us and if you stay within that broad front you are more likely to do sensible things whereas if you go off on some specialised little track of your own, dreaming up some axioms and pursuing them, you probably will end up doing nothing worthy ...Being within the grand themes is a good test of worthwhileness.*

Two pure mathematicians had surprising perspectives, the first invoking story-telling as a description of what mathematicians do and the second being critical of the formal tradition of mathematical writing and claiming that mathematics is what you understand, not what you prove:

> *The only thing mathematicians can do is tell a good story but those stories do uncover mathematical truths - mathematical truths are discovered. However, the mathematics that we know, write down, is made up by men and women. Statements, which have, in today's society, been accepted as proofs, under the conditions of today's society, are what I am calling mathematical truths.*

> *The definition, theorem, proof style is sometimes necessary to the health of mathematics but it can be over-prescriptive. People think that is what maths is whereas I think it is about filling in gaps, making the map. Maths isn't what ends up on the page. Maths is what happens in your head. I don't think maths is about proving theorems. It is one constituent but maths is about mapping abstract ideas in your head and understanding how things relate.*

A difference between pure and applied was captured by this applied mathematician:

> *I see mathematics as a way of describing physical things using formulae that work, modelling the real world and having a set of rules, a language, which allows you to do that.*

The existence of the kind of diversity represented in the above quotes led me to expect that a range of different beliefs existed across the discipline as a whole. I have pointed out in Chapter 2 that these beliefs are held within a philosophical position, which Sal Restivo has called "transcendental", although, at the same time, he pointed out that even the transcendental is a human construction:

> Truth and knowledge, as fallible and tentative achievements, are manufactured by human beings who accomplish what they know and what they can know in common. (1999:4)

As Philip Davis and Reuben Hersh (1983) demonstrated, mathematicians are past experts at shifting their positions:

> the typical working mathematician is a Platonist on weekdays and a formalist on Sundays. That is, when he is doing mathematics he is convinced that he is dealing with an objective reality whose properties he is attempting to determine. But then, when challenged to give a philosophical account of this reality, he finds it easiest to pretend that he does not believe in it after all. (p.321)

The mathematicians that I interviewed, in fact, demonstrated the differences described but, if anything, the other way around! Their convictions of an 'objective reality' were most usually connected with the results or outcomes of their research whereas some of them appeared to feel more comfortable with denial of 'objectivity' when describing their research process. For many, however, *My success as a mathematician rests on the fact that I don't ask too many questions about how it happens, how I got there. I don't think about it, I am just curious and I do it.* At the same time, they appeared to keep a tight hold on the absolute nature of 'truth' and the power and legitimacy of the objects of mathematics, whether 'discovered' platonistically, or derived through game playing of a formalist kind. Interestingly, none of them spoke specifically about mathematics being the study of patterns and relations (see Sawyer, 1943). They conformed to the picture painted by Robert Thomas (2000):

> I think that both mathematicians and philosophers have been deceived to a greater extent than they would normally allow into regarding mathematics as being about mathematical objects...The reason that I think the subject of the argument [the reality or otherwise of these supposed objects] does not matter is that I think mathematics is about relations rather than objects...this claim ...is often remarked on but then ignored. (pp. 301-2)

Diversity was also exemplified on other dimensions of the model. So, whether or not a mathematician held absolutist views about the discipline, they each positioned themselves on the role of aesthetics and intuition along a continuum from unimportant to crucial (see Chapter 5). Only when it came to speaking of connectivities within and without mathematics was there almost universal agreement as to their importance, whether or not their current work could be so connected (also see Chapter 6).

> *Sometimes one comes across the missing link. A small idea closes up a circuit that has been around not closed up for a long time.*

> *Your work is not geared immediately to applications although it would be excellent if one could establish the bridgehead which allowed that.*

Some ideas that were understood by one group of people were exactly what another group of people needed to make something work.

You feel that what you have been doing is part of something bigger.

However, diversity does not appear to affect the wider, societal view on mathematics, most especially the view that influences how mathematics is understood in schools. Indeed, the epistemological diversity to which I am drawing attention co-exists with a philosophical uniformity, a strong commitment to The Big Picture in mathematics together with an image of an individual's contribution of one more piece to the mathematical 'jigsaw puzzle'. (Newell and Swan (1999) pointed out that the jigsaw metaphor is connected to a 'traditional', positivist, approach.) This connection between the positivism of the mathematicians and their expressed preference for the jigsaw metaphor as a description of coming to know, is very visible in the data.

Newell and Swan (1999) called for innovation through knowledge sharing and dispersal. Karin Knorr-Cetina (1981) made plain that innovation, despite many stories to the contrary, is far from accidental and might, perhaps, be connected by mathematicians and scientists to their "jigsaw" conception of knowledge.

> Innovations are the result of intentional, directed work, and not merely chance happenings. It is the scientists' knowledge of what is a problem and what counts as a solution, educated guesses about where to look and what to ignore, and highly selective, expectation-based tinkering with the material that guides them toward an 'innovative' result. Once a result has been obtained, the careful selection of a publisher (and therefore, an audience) as well as various marketing strategies, can turn a laboratory product into something which may be widely accepted as 'new'. Nor should we forget that, to a great degree, scientists select areas of work which have not been covered by previous research; thus, their results are almost guaranteed to pass as new...In short, there is nothing non-directional or purely random about the individual scientist's efforts toward innovation. (pp. 12/13)

Many of these comments apply to the work as described by the mathematicians.

Nonetheless, it is very encouraging that the uniformity, which so often features in the public pronouncements of mathematicians, or of non-mathematicians, about mathematics, is not matched by the voices of my participants as they described their experiences. Furthermore, the spread of positions raised, for me, the importance of making dialogue about the discipline a necessary feature of its learning. I will return to this in subsequent chapters.

SUMMARY

In this chapter, I have explored the 'histories' of the mathematicians, as explained by them, in order to challenge stereotypes and raise interesting or concerning issues. The notion that mathematicians emerge at birth from mathematically connected families is not sustainable. Many of these mathematicians, while expressing interest in mathematics at school, also expressed interest in other areas, including languages and the arts. To some degree, it was the influence of a particular teacher that supported the choice of mathematics. This was particularly important for those, of whom there was a surprisingly high number, from families where there had been no experience of university. In the case of a few females, although the person concerned was already succeeding in the discipline, it was more luck than anything else that caused them to enter a career as a mathematician. What was noticeable was how lacking in information, at the moment of choice, were many of the mathematicians (and see Zevenbergen, 2003 for an application to teachers' beliefs in the mathematics classroom).

Being an only child did not seem to be connected to career choice in any way but 50% of the participants were eldest children in their families. This seems greater than might have been expected, but whether it was causally related to their choice of discipline cannot be said and was not related to their sex. Differences in career paths were identified, in particular between those paths that appeared to be smooth, and those that appeared much less so. Women, but no men, mentioned career breaks. The role of Oxbridge in the lives of these mathematicians was very important and what they said, particularly the women, about their experiences there, even more so. Half of the 28.6% who mentioned family connections with mathematics named engineering as being the connection via father, grandfather and brothers.

Great heterogeneity, unrelated to gender, was observed in the beliefs that were expressed about the nature of mathematics; however, these were couched in a contradictory, fairly homogeneous and predominantly Platonist version of mathematical knowledge. This contradiction between heterogeneity of position and homogeneity of beliefs is further explored in Chapter 10.

In Chapter 2, I presented the epistemological model, and, in this chapter, provided background information on the mathematicians; this has displayed the diversity within this community. In Chapters 4 and 5 I turn to three aspects of the model to explore this heterogeneity; first, in Chapter 4, I deal with thinking styles and then, in Chapter 5, aesthetics and intuition and the feelings that the mathematicians expressed more generally. These 3 categories of the model deserve particular attention, I believe, because of the surprising results of asking the mathematicians to explain how they think about the problems on which they are working. But also we hear the voices of the mathematicians as they blend the affective with the cognitive in ways that are particularly appropriate to learners.

4

How do mathematicians think about mathematics?

At the outset of this study, I had two conjectures with respect to thinking style. The first was that I would find the two different thinking styles, the visual and the analytic, well recorded in the literature (for example, see Hadamard, 1945: 86). The second was that research mathematicians would move flexibly between the two. Neither of these conjectures was confirmed.

MATHEMATICAL PROBLEMS

I approached the issue of thinking styles by asking the mathematicians to talk to me about the problem or problems on which they were currently working. I asked on how many problems they usually worked in the same period. One mathematician made the comment with which many would have agreed: *It rarely happens that one thinks I would like to know this, let's go out and find the answer. It is much more that there are various problems that one is aware of and something comes along and you think of how that fits.* Their responses indicated a range from one problem to "lots", the majority falling between two and five. Most remarked that they *usually work on too many problems at one time.* One said: *I have twenty or so problems on the go, all in my head, and you pull a different one to the front or it pops to the front.* Another commented: *I am really only thinking about two or three problems but one of those is a really big problem which I have been working on for twenty years and I have recently made just a bit of progress.*

Where does the thinking happen?

Next I asked them where they did their thinking work that is in the absence of recording devices such as pencil and paper, computer or board. They gave me variants on the following although the elaboration of "in bed", below, was unique to one mathematician:

> *I do think walking so when I am stuck I do go for a walk;*

> *I typically solve problems in the bath;*

> *When I am thinking about a problem, I have an infinite board in front of my eyes and I draw pictures on it. It is about 2 feet before my head, and I write on it with my finger. I read it, do calculations. I find it very difficult to do it without using my finger. That is the way I generally do work ...typically I do it at night in bed. I think, principally I imagine, algebra and write down the equations. I do some sort of calculation. I could quite easily go on for five hours doing this in my head.*

And perhaps most worrying was:

> *I do all my thinking in the car.*

Aeroplanes featured in a few of the stories: *My Australian colleague proves theorems on aeroplanes in his head.*

Styles of Thinking

I then asked the mathematicians to focus on a problem of their choice and to describe to me what was on their "mental screen". Some expressed concerns that I might not understand so I developed a strategy of listening and then describing to them what I was "hearing" and this appeared to satisfy them. Indeed, some complimented me on how clearly I could describe what I had heard, which I put down to being a pedagogue, rather than a mathematician. Other than that, they did not seem to have a problem with my request.

When we had thoroughly explored their chosen problem, I asked if they thought about all their problems in a similar way or, if they chose another problem, would that evoke a different way of thinking. Twenty-five claimed always to think about all their problems in the one way described, thus falsifying my second conjecture that all would move flexibly between different styles of thinking. The remainder said that their thinking style was problem-dependent so I asked them to choose another unsolved problem which, preferably, demanded a different approach and we went through the same procedure.

When talking me through how they solved problems, I found that the mathematicians offered evidence on three, not two, different thinking styles thus falsifying my first conjecture. I call these three styles:

Style A: Visual (or thinking in pictures, often dynamic),
Style B: Analytic (or thinking symbolically, formalistically) and
Style C: Conceptual (thinking in ideas, classifying).

Differentiated between pure and applied mathematics and statistics, my participants spread across the three styles in the following ways (all bracketed numbers show female/male):

Table 4.1: Thinking style by area of mathematics

	Pure	Applied	Statistics	Totals
Style A	21	15	9	45
	(11/10)	(5/10)	(5/4)	(21/24)
Style B	11	11	6	28
	(7/4)	(5/6)	(2/4)	(14/14)
Style C	10	16	7	33
	(6/4)	(6/10)	(4/3)	(16/17)
Totals	42	42	22	106
	(24/18)	16/26)	(11/11)	(51/55)

The numbers do not sum to 70 since 39 mathematicians claimed to make use of more than one style. Of those 25 who used one style only, for 15 it was Style A (9/6), for 3 it was Style B (2/1) and for 7 it was Style C (4/3). In Table 5.1, Style A, visualising or thinking in pictures, dominates particularly for pure mathematicians but the distribution for applied mathematicians and statisticians is more evenly spread. Of the 70 mathematicians, only 3, 2 females and 1 male, appeared to use all 3 styles, whereas for 36, it was a combination of 2 out of the 3. Six mathematicians are not included in these figures because of insufficient information.

A number of mathematicians made similar categorical statements to the following: *I am not a visual person. I do not develop things in terms of pictures. I am profoundly unvisual* and, in contrast: *If you can do things with pictures I am allright but I start getting lost if I cannot picture it.* There is a generally held view that visualising is integral to being an active mathematician and I was given an impression by some of the mathematicians who did not think visually that, in the hierarchy of mathematicians, not to think visually was a little less acceptable. To coin a phrase: Proper mathematicians do it visually! From my data, as can be seen in Table 5.1 above, there is nothing to support this.

Those who claimed not to use a visual style of mathematical thinking appeared, like the mathematician quoted above, to be aware that there are alternative ways of thinking about mathematics but many of the others appeared to hold an assumption that everyone thought about mathematics in the same way that they did. Even if one accepts Style 3 thinking as a form of visualization through classificatory mechanisms, there is still a significant roughly 26% (28/106) using a non-visual thinking style (Style B) for some of their problem solving; 12% (3/25) were single Style B users, approximately 5% of the total number on which these data were collected (3/64). If the Style C thinkers are included in this calculation of non-visualisers, well over 50% of the cited styles were non-visual (61/106) when using Styles B and C. This does not mean that they are not visualizing something on their mental screen but, in their own terms, they are not visualizing mathematics.

What did the mathematicians say that caused me to classify them in these ways? Some visual thinkers said:

I largely think of it as a picture. A lot of things I think about are represented as pictures.

Mostly I think pictorially. I see things moving through walls, the way mirror images translate into this different picture where things move through walls.

I think in pictures. I have a kind of photographic memory. I see colourful pictures; they have to have colour in them.

When I am listening to papers at conferences, if they provoke images in my head it makes it easier, but it isn't the only way of thinking about a problem.

Sometimes pictures tell you a lot and I can see things in mathematical pictures very clearly but it is not the first way I would think.

Some analytic thinkers said:

Sometimes I probably am just thinking in the symbols representing the mathematics that I am doing and those symbols probably become the words.

I probably think in terms of a sequence of distribution shapes or actual probability functions or something like that. I am trying to visualize what would happen if I took the ratios of two normals or something. I can quite easily visualize the functions and the equation systems and I can do a certain amount of manipulation of that. I would see the algebraic formulae and I would manipulate those formulae in my head.

I think in formulae or words, more likely to be formulae. Sometimes I manipulate them, sometimes I try to work out what the sensible next move is.

I could see a spreadsheet of data, a couple of matrices and I was trying to manipulate these matrices and formulate the constraints in symbols, vague symbols.

When I think about a problem, I see it in my mind as equations...I manipulate these equations on my mental screen.

Some conceptual thinkers said:

I don't see equations and I don't see maths in pictures. With me, it is about approaches.

It is not pictures, it is more ideas...I have a conceptual picture of what is going on and try to clarify it...I might classify the ideas, combine or disaggregate them.

It is quite word dependent. So I have words in the boxes and connectors between the steps and I might be interrogating the words or the connectors.

If I thought about a different problem, I would probably see blobs, sets of mathematical objects, a fairly visual geometric thing, pictures of things, rather than words or equations. Not a mathematical picture; ideas, putting them together, playing with concepts.

So there is a space in my head that has ideas that I am working with positioned and I can play with those ideas, re-position them, relate them, and so on. It is certainly partly visual but maybe conceptual is a better word.

Examples of the 3 possibilities of two style combinations are:

Styles A/B: *If you are doing theoretical physics you are constantly flipping backwards and forwards between formal manipulation and the visualization of something and the role each of these things plays is not very easy to pin down...So I am relating the formalizations to the pictures in a fairly fluid way and I am moving between them.*

Styles A/C: *I always see the set rather than the group that acts on it. It is a mental picture. I cannot imagine thinking in a way that is completely non-visual...Depending on what the problem is but usually I am trying to count how many different orbits there are so things fall into classes so you can move within a class.*

Style B/C: *I am not a very visual person. I think in terms of equations but I think it is problem driven. I don't think I can ever deal with solid geometry problems. If I am looking for a taxonomy, I use hierarchies, classificatory pictures.*

One of the three participants who used all three styles said:

Pictures, I think in mathematical pictures ...I think there is a third mode, neither spatial nor analytic...With certain things that I do, you are considering things which have a structure, n-dimensional space say. Sometimes you have equations and sometimes I don't know how I do it. Sometimes, which is more overall, you have the mathematical concepts. If I think of something being a sub-set of something, I am thinking about concepts.

Out of this classification system, remembering that I coded 106 different responses to the question about thinking style, 45/106, under 50% were classified Style A, 28/106, just over 25% Style B and 33/106, over 30%, Style C, each alone or in combination. The breakdown between female and male does not justify an assertion of difference and certainly does not substantiate the stereotype of the visual male (see Table 4.1).

An interesting feature of the data was the relationship between thinking and writing. Brian Rotman wrote: "mathematicians at the same time think their scribbles and scribble their thoughts" (2000: 35). When the mathematicians were thinking not all of them wrote, as in scribbling on paper, but some, like the mathematician quoted early in this chapter, did notate on their 'mental screens'. He added: *I get very annoyed at having to write things down on paper. It's slow, cumbersome, you can't do it anywhere.* In contrast, another said: *The second I have the idea, I cannot think any more without writing it down. It establishes the conditions and the constraints on the formalism and its boundaries.*

They were searching for a trigger to help them make connections. These triggers arrived from many different sources.

Writing down what I can do helps me to walk away from it. Lying in bed at night, thinking I could try this or that; I write a short list of what I could try in order to help me switch off. Then I come back to the list and look at it and it might set me off on a new direction.

I might go and read journal papers, math reviews...I might formulate a question for a collaborator. That's marvelous when it works.

> *I keep going to the Library and trying to find something in the literature. Sometimes I talk to people*

Seminars are often a productive trigger, for example:

> *What he was talking about was a simple case of what I was trying to generalise. Twelve hours later I got on the plane, got to Australia, got up in the middle of the night and wrote it all out and it was perfect. It wasn't an idea that was missing, it was the notation.*

Stuck!

Fifty-five of the mathematicians referred to the state of being stuck. One said: *The natural condition of doing research is to be stuck, most of the time, on most of the things you are doing.* Another said: *When I am stuck, I hope I have learnt to welcome stuckness almost as the natural state of a mathematician.* So what did they do when stuck? The majority (18 mathematicians) spoke about *walking away,* or *giving up for a while, letting things vegetate, doing something quite different,* or *moving to another problem.* You might distance yourself from one of the states of being stuck by walking around a bit or having a cup of coffee: *When I am stuck, my strategy is to catch up on all the other things I ought to be doing, and then go back to it.* The importance of having a number of problems on the go at once, was that: *If you have enough things to move to when you are stuck, you leave them alone and walk away to something else.* Divergence, again, was demonstrated by two mathematicians, one of whom said: *I give up fairly easily, walk away, and start again later, if I am stuck,* while another said:

> *If I am stuck, I am extremely stubborn and can continue for weeks so it is a long time until I actually concede that I am truly stuck. But then I would walk away!*

Eleven mathematicians mentioned, when stuck, going to talk to someone and a further twelve read books or journals. Five specifically mentioned, when stuck, using the computer. Both being stuck as a natural state and walking away from it, as a reasonable thing to do, carry implications for students that are addressed in Chapter 10.

I am reminded of the chapter we wrote in *Thinking Mathematically* (Mason, Burton & Stacey, 1982) that we called Responses to being STUCK (pp. 49-61) in which we said: "Everyone gets stuck. It cannot be avoided, and it should not be hidden. It is an honourable and positive state, from which much can be learned" (p.49). In a later chapter, Still STUCK? (pp.105-114), we drew attention to the strategies that can be invoked to help but also to the unhelpful mind-set that can be induced by holding hidden, or unnecessary, assumptions. The mathematicians also had many strategies for dealing with a state of being stuck.

> *If I am stuck because the calculation is too long and hard to do, you have either to decide that you are going to work hard and long or that you have to find another way of doing the problem, or give up. But many kinds of stuck, you simply go away and do something else and hope that something turns up.*

For another: *It is very easy to get stuck in one way and you need to get further away and see what that is telling you. I do think walking so when I am stuck I do go for a walk.*

It is not easy to convince students that, when doing mathematics problems, it is an honourable state to be stuck. However, the mathematicians clearly believed this, and had ways of recognising and dealing with the state. I am convinced by their strategies and I think it is a pity that students do not get the opportunity to share the state, to experience its positive side and explore ways out of it. One mathematician summed up:

> *When I am stuck, I do find it very frustrating. I know it is normal and a natural part of mathematics but I do worry that I won't get out. I try to read around the problem, find some book that might have some relevance, try to isolate what is the real problem. When you find out that you are wrong, it isn't that bad because when you find out that you were mistaken it is because you have found the real answer. The relief of getting over the stuck part, it doesn't matter if the answer isn't actually right. You have broken through and even if it doesn't work, there are other choices.*

CONCLUSIONS

So, what conclusions can I draw from the falsification of my two starting conjectures, and the derivation of the three-way classification scheme for thinking styles in mathematics? First, to reiterate the two startling conclusions:

1. There were 3 different thinking styles evidenced in what the mathematicians told me about how they thought about their problems.

2. Very few of the mathematicians used all three styles and some only used one style to think about the problems on which they worked.

What consequences do these conclusions have for mathematical classrooms? It is unclear to what degree thinking and learning styles interact but one cannot be independent of the other. Most of my participants told me that they had not thought about these issues and, certainly, they were not aware of the work done by John Cowan in engineering education. John Cowan is the only person of whom I am aware whose research focused on differences in thinking style in both teaching staff and students and their implications for learning. He identified three different thinking styles amongst structural mechanical engineers. "Almost everyone used each strategy at some time or other-but most had a preferred strategy they tended to favour, even for problems demonstrably more suited to a different approach" (Cowan: 1983: 303). This is similar enough to the mathematicians to raise a sense of confidence in my data. John Cowan, speaking of both staff and students, went on to say that:

> Most of my subjects in extended recorded protocol inquiries have become so interested in the research that the whole activity had a productive tutorial effect on them. They are now more aware of what they are doing, can proceed more effectively, and have become better problem-solvers in many disciplines. (Ibid: 303)

He did not speak about the impact on their students. However, he developed self-study packages covering similar material but utilizing the approach of each of the different thinking styles; he recommended to students that when they had used the package that reflected their style and resolved their difficulties, they compare the contents of that package with the other two. He claimed that these approaches had very positive effects on student learning (see Cowan, 1975).

The results reported in this chapter probably carry the most serious implications for the teaching and learning of mathematics arising from the study. Instead of a teacher viewing a student who fails to learn mathematics as having a deficit, the non-consistency of thinking style could explain many failures in communication between teacher and taught. Responsibility, therefore, shifts to teachers to be reflective about their own preferences as well as instrumental in making their students aware of different thinking styles and how best to make use of their preferred style and compensate for any deficiencies. In Chapter 10, these implications are explored along with dealing the natural state of being stuck.

SUMMARY

Three different styles of thinking about mathematics were identified. These were labelled as visualisers, analysers and classifiers. Fewer than 50% of the total responses were coded as using a visual style of thinking. The remainder were distributed between the other two styles roughly equally. From an educational point of view, the significance of the outcome on thinking styles connects, as it must, to how we learn. What impact on learning would a raised awareness of thinking style have on teachers and students? This question is addressed in Chapter 10.

In Chapter 5, feelings about engagement with mathematics take centre stage. First, two categories of the model, the role of aesthetics and the presence and use of intuition or insight are discussed by the mathematicians. The chapter concludes with a consideration of feelings and the role they play, for the mathematicians, in being producers of mathematics.

5

Aesthetics, Intuition/Insight[1] and the feelings associated with mathematics

> *I am not talking about the fact that the diagrams looked pretty on the board. I am talking about the sort of excitement that you might get from looking at something structurally beautiful such as a beautiful picture or hearing a particular piece of music that was rendered particularly beautifully by a certain artist. It is the chain of logic, the structure connecting the beginning to the end, the completeness, the structure of the proof appealed to me aesthetically. When I chose a branch of mathematics it had to appeal to me in that particular kind of way.*

In this chapter, I explore the role the mathematicians perceived for aesthetics in their coming to know but also look more closely at intuition (which some mathematicians preferred to call insight) and the connections the mathematicians made between intuition and aesthetics, *intuition and beauty are related.* The mathematicians discussed aesthetics and intuition/insight in terms that were emotive, full of expressed feelings. Both appeared to contribute, sometimes severally, sometimes together, to the ways in which the mathematicians came to know. Their role is, therefore, very interesting, spanning and linking the domains of feeling and knowing. Unlike most of the mathematicians, including the one whose quote opens this chapter, the aesthetic described by the next participant lies not in the product, that is the proof, but in the chosen strategy to reach that end:

> *How you choose to make the decisions about the route you are going to follow constitutes the aesthetics of mathematics. I make these decisions by guessing – that is the intuition part – it feels good.*

---

[1] A first version of part of this chapter originally appeared as the article Why is Intuition so Important to Mathematicians but Missing from Mathematics Education? in *For the Learning of Mathematics*, **19**(3) November 1999, 27-32.

I return specifically to the feelings that the mathematicians expressed about their subject in the final section of this chapter.

As I have indicated in other chapters, I do not believe that psychological, or genetic, explanations, frequently found in the literature, are adequate to explain links, in this case between aesthetics and intuition. We learn to recognize what others consider beautiful, elegant or ugly and, as we are inducted into the mathematical community, to identify with and use such descriptions and to have our own impact on them. Of course as I make clear below, the personal feelings that the mathematicians report, associated with their research, are important to their motivation and commitment. But, for explanations as to their prevalence and the descriptive similarities, I am looking to the community for the ways in which such feelings are defined, acknowledged, reinforced or denied.

AESTHETICS

Forty-three mathematicians, just over 61%, introduced aesthetics, beauty or elegance into our discussions; the majority spoke of beauty. This high number is not surprising given the frequency with which aesthetics feature in the discourse of mathematicians, especially those in pure mathematics. Ian Stewart wrote: "Patterns possess utility as well as beauty" (1995:3). Reuben Hersh (1998: 33) called elegance "desirable", not "essential" with which I believe many of the mathematicians, especially those in applied areas, would agree. Douglas Hofstadter (1979: 584) linked "beauty" with "truth" as both being "intangible notions". Of the 43 mathematicians, 3 made negative comments either about the relationship between beauty and mathematics: *Beauty doesn't matter. I have never seen a beautiful mathematical paper in my life*, or about the personal nature of aesthetics: *I am not sure that there are things that everybody would agree are beautiful*. However, this latter mathematician, when speaking of the review process for papers, and how she felt about rigour of argument, did say: *If it was a very elegant way of doing things, I would be inclined to forgive a lot of faults.*

Differences in how they approached aesthetics, amongst other aspects, could sometimes be tracked to whether they were pure or applied mathematicians although this wasn't always the case. The first mathematician above was a numerical analyst, the second an applied probabilist. However, while most pure mathematicians appeared to consider beauty as intrinsic to mathematics, applied mathematicians linked aesthetics with pure mathematics, as one explained: *Applied mathematicians probably dispense with the elegance simply to get a mathematical solution which works.* Another applied mathematician said:

> *Elegance is important but that doesn't completely square with my ethos as an applied mathematician. There is a compromise in being a pure mathematician driven by elegance and an applied mathematician driven by the problem.*

A pure mathematician who had *moved to applicable mathematics but am not myself an applied mathematician* said: *One of the things I don't like about so-called applied maths is that they don't seem to worry about getting the right answer.* Another pure mathematician gave his perspective:

> *Particularly in this country, there is a big gap between pure mathematics where things are done very carefully and applied mathematics where they don't care about things like that.*

I found many differences between the communities of pure and applied mathematicians, of the kind displayed around aesthetics. In some universities, the 2 communities are housed completely separately and this is despite such categories being neither robust nor prevalent even across the UK, never mind internationally. For example, some of those working on the boundaries of physics and mathematics pointed to differences between universities that dictated whether you were located in a mathematics or a physics department. A participant explained:

> *The separation between pure and applied in this country is not reflected in other countries. For example, in Germany what I am doing is regarded as functional analysis. Their applied mathematics is almost like what we would call engineering. The divisions are not in the same places in different countries.*

Although these designations are somewhat hazy and culture-centric, I maintained them in order to protect the identity of my participants. But also, many of the mathematicians categorised themselves as in pure, applied or statistics and, with the category, came the kind of cultural baggage, including the 'us-them' approach, already observed above (and see Burton, 1999b). So, ways of understanding aesthetics that, for pure mathematicians, were different from those for applied mathematicians, were not unexpected.

To comprehend the role of the aesthetic in mathematical thinking, it is important first to appreciate the research participant's location and conception of aesthetics. I have already explained that aesthetics appeared to feature more in the discourse of pure mathematicians than applied or statisticians. This was because, for the majority, the aesthetics were connected to the product, the proof. A very few mathematicians, often applied, spoke of the aesthetics of process or what, below, I call their generative function. To explore the mathematicians conception of aesthetics, I used Nathalie Sinclair's (2002) tripartite model of the characteristics of mathematicians' construction of aesthetics, based on mathematicians' descriptions of aesthetics as found in the literature (and see Chapter 6). She proposed three characteristics of aesthetics: the generative, the evaluative, and the motivational. The generative focuses on the role of aesthetics in inventing or discovering mathematics; the evaluative typically manifests itself in the judgement of a mathematical product, such as a theorem; the motivational relates to the role of aesthetics in prompting or inspiring mathematical activity. Another equally important dimension to a consideration of an epistemology based on aesthetics must be that of function: how does the (or an) aesthetic operate as a way of knowing?

In analysing what the mathematicians said about aesthetics, I tried to separate their comments according to whether they seemed to be generative, evaluative, motivational or were describing how something functioned in contributing to their knowing. I deal with each in turn but by far the greatest number of comments related to function. Nonetheless, the beauty perceived in structure connects beginning to end and hence can be identified as both generational and evaluative as well as providing functional categories. (For example, see the quote that introduces this chapter). Similarly, where such functional comments are quoted in this chapter, an aesthetic characteristic may be inferred. From the three aesthetic characteristics, it was the evaluative that drew more comments than the other two, reinforcing the notion that judgments about beauty are most frequently related to beauty of the product, i.e. proof.

Aesthetics – Generative

Only two mathematicians made comments that could be directly construed as indicating that aesthetic qualities provoked them further in the process of research. One said:

> *I have just read a paper and a more recent paper. In the older paper they constructed something without explaining why it worked. You know for meta-reasons that something should exist but proving it is a matter of calculation. Then later, someone comes along with an explanation for why it exists. The first paper shows that it is true; the second paper explains it. The explanation is the beautiful thing.*

Although the mathematician speaking was not the one to move from constructing to explaining, it seemed to me that he was indicating that construction without explanation provoked him towards generating new mathematics. I classified this comment, therefore, and the one below, as both generative and motivational (see below). The second mathematician used the adjective 'nice' and was not the only person to do so.

> *We have to prove nice theorems. We have some view of what it is that makes a nice theorem. That is an aesthetic judgment. Firstly it shouldn't be too complicated to state. Theorems that require three pages of hypotheses are not nice theorems. But in some sense it should open your eyes to something that you haven't noticed before. You should read a nice theorem and say: oh yes, it must be so. It shouldn't be accidental or be proved because somebody battered it to death.*

In this case, I interpreted the mathematician as saying that, evaluatively, "nice" means uncomplicated and, in the best way, obvious; one might say, retrospectively fitting, or connected. But I detected a flavour of the generative in that I would expect that this mathematician, faced with a set of compact, "nice", conjectures, would be motivated towards generating a proof because of an expectation that the proof would substantiate the existence of *something that you haven't noticed before*.

Aesthetics – Motivational

Brian Rotman (2000) pointed out that because proof is about convincing, oneself and others, it involves a "process of persuasion" (p.17 et seq.). Attempts to convince are motivated by "the story that is being told, the narrative through which the thought experiment or argument is organized" (ibid: 18). When it came to identifying the motivational aspects of aesthetics, two more mathematicians, in addition to the two above, made statements that seemed to me to reflect the ways in which aesthetics acted as a spur to further mathematical activity. One, when interviewed, had, after considerable thought, ceased a research career in mathematics but commented:

> *When I decided to leave mathematics, the one thing that would keep me there would be these five seconds, these moments when you feel you have seen one of the most beautiful things in the world.*

The second was the only mathematician to refer to the difficulties associated with trying to turn a proof into an elegant proof. He said:

Elegance requires constant refinement, a lot of work. I think the easiest way is to write a draft and leave it for sufficiently long a time that you have emptied your mind of that material so you have pared away all the stuff that was in your mind when you wrote it. That is a very time-consuming process. So elegance is about being uncluttered with irrelevance. Quite often you need to work on a proof to eliminate steps which are not serving any useful purpose.

There are interesting similarities between this description and those made by artists and sculptors but the relationship between aesthetics and rigour that, in mathematics, stresses rigour, in the arts stresses aesthetics.

However, two more mathematicians spoke about the negative side of 'beauty', which one referred to as ugliness but also made clear that ugliness and beauty were intimately connected and subjectively decided. And having to use 'ugly' methods did not de-motivate:

Beauty and simplicity tend to go together but it is an unfortunate fact of life that the first time you do something it can look terribly ugly just because you are not used to it, or you may have found a bad way of doing something which can be done more easily another way. It is very satisfying to do things which are beautiful but I don't think that beauty is a criterion for whether or not something is worth doing. If one's treatment of something is ugly, too bad, it is an unfortunate fact of life. String Theory is something I can write down in about two lines and it looks very beautiful to me. It is true that I cannot explain its consequences without doing unpleasant things. That doesn't make it ugly to me but it does to some other people.

The other mathematician to make a similar point reflected on her feelings without giving any explanations for how she differentiated between "nice" and "nasty":

There are nice proofs and nasty proofs but in differential equations, at the moment, every so often you have to do hard analysis and it might be clever what people do but it looks hideous on the page.

Aesthetic - Evaluative

The comments that were evaluative also, of course, referred to functional categories. So, in the next quote, in evaluating elegance, the mathematician used the functional category of "succinct" as a way of deciding if a proof was elegant. However, this also provided a functional category for 'knowing'. Interestingly, again, this applied mathematician made the point of a difference between applied and pure mathematics where elegance is integral:

> *At that time Lighthill was producing some very good work. Lighthill's papers were such elegant pieces of mathematics. So many problems nowadays are computerised in a way that, for me, some of the excitement has gone. All the work that I have done has been very analytical. I like analytical solutions that are succinct, use classical mathematics, you don't need to make too many approximations. Elegance has nothing to do with applications.*

But, another mathematician disputed this point: *If someone finds a result which is true, with an interesting application, odds on it will be beautiful anyway.* Succinctness featured again in the following:

> *If it involves too many complicated, messy expressions I don't find it very beautiful or elegant. There are whole parts of particle theory which I never managed to get to grips with because they seemed so messy and complicated. If something can be expressed in a compact notation it adds to its elegance.*

But not for everybody was elegance necessarily a function of length or succinctness: *It is the ideas being elegant rather than the paper being short or concise. So the ideas should be general.*

To succinctness, some mathematicians added structure: *I think that elegance is more precise. It has to do with structure.* Structure evoked strong aesthetic feelings:

> *I suppose one gets a sense of exhilaration, an appreciation of its elegance - a fit of form and shape in a very abstract sense just as in certain paintings or music you have got a structure which holds together with a sense of both differentiation and balance and wholeness. Something like that - something like perfection. I don't have a problem with imaginative perfection.*

More obviously evaluative was the comment connecting complexity in the process of proving with the beauty of the argument:

> *Some of the nicest problems in mathematics can be stated completely simply but then when you look at the bag of tricks that is involved to prove it you get this huge great machinery that has its own rather beautiful structures. There is that mathematical aesthetic.*

The same mathematician invoked one of the categories of the model, that of connections, as yet a different evaluative aesthetic:

There is also a physical aesthetic where you are actually saying something concrete about, say, Einstein's Theory of Relativity.

The link between mathematics and the visual is often asserted but some mathematicians made clear that while visual sometimes means seeing: *If there is symmetry, it is pleasing to the eye,* sometimes seeing acts as a metaphor for what is in the mind's eye:

Beauty is very personal although there is a surprising amount of consistency. It is very hard to explain. I find a level of beauty in wooden puzzles. If things fit together nicely, in a smooth way, no gaps and it feels nice...it is that sort of idea. A number of ideas just fit together nicely with no gaps.

As with the motivational characteristic, there were mathematicians who spoke of aspects of mathematics that led to negative feelings. First the technique of simulation: *There is something not very satisfactory about simulating.* For the next mathematician, utility and aesthetics were not connected. Neither did being correct necessarily make something aesthetically pleasing:

Aesthetic qualities are definitely there. Sometimes if you are being a referee you might recognise the work that somebody has done and it is perfectly correct but we aren't interested because it isn't aesthetic.

And the next mathematician knitted the aesthetics into the epistemology:

I would hope that it had a sense of beauty but if it didn't I would just be disappointed. I think beauty in maths is that it gives me a feeling of satisfaction, makes me feel, oh, that is nice, or that is really interesting, or completes another bit of my knowledge or smooths over a hole in my knowledge. I think in a way even if I had never thought of asking that question, there is a question that I could have asked that now I know the answer to. So knowledge is a question and an answer.

Function

How does the aesthetic operate as a way of knowing? Above, I have already identified a number of functional categories all of which relate to qualities of proof. These were succinctness, generality, structure and connections: *Unusual connections would be beautiful.* Instead of succinct, the word 'concise' or 'compact' was often used as by the next mathematician. He also offered novelty, power, rigour and deep implications as additional categories.

> *To be elegant, a piece of work has to be a very advanced body of work, using quite novel techniques and use them in a powerful way. The implications for the statements should be quite phenomenal. You solve a difficult problem in a compact way, completely rigorous - simple statements which have fantastic implications.*

Simplicity was a functional category mentioned by many. For example: *A simple statement that is also general, that is the tops for elegance.* Individual mathematicians raised a number of other functional categories, including accessibility, interest, satisfaction, symmetry, transparency and surprise, reference to many of which can be found within the quotes above. However, as Candia Morgan pointed out, such judgments rest upon:

> mathematical claims, arising from a deep enculturation into mathematical ways of thinking, resting on a conception of mathematics that assumes regularity, connectedness and the legitimacy of using analogy to discover (or construct) new types of object. The claims would not make sense to a reader who did not share this enculturation. (2001:174)

Whereas the writing addresses the mathematical claims, the aesthetic judgments make use of these functional categories to influence how these claims are evaluated. The power of the enculturation is thereby substantiated.

Curiously, the mathematicians did not speak about the role of the imagination and it was not something that I raised with them. But Robert Thomas wrote "in mathematics one imagines in order to see why what is implied is implied" (2002: 45). He continued:

> the function of much mathematics learning is to stock the imagination, not wholly differently from the way learning stories stocks the imagination. One sees what may be and – unlike stories – what cannot be. As well as learning to see in mathematical (relational) terms, one also learns that some relations are not possible in the presence of others. (Ibid: 45)

I think this comes very close to applying what Stephen Brown and Marion Walter (1990) called their "What-If-Not" Strategy, a process which, I believe, they justifiably claim is not only creative, "enables us to uncover unsuspected depth in starting points that may appear pedestrian" but also "[p]roblems that have been unsolved for centuries in one context reveal themselves with ease in others" (p.102) This is one way of understanding the functional value of overtly making use of the imagination to stimulate mathematical creativity.

To summarise, aesthetics are, for most, a product both of the mathematical culture and of the culture of mathematics. Within the culture of mathematics, we learn to recognise what, for the community, constitutes aesthetics: structure, compactness, connections or any of the other functional categories that help to define aesthetics as a way of knowing, particularly in relation to the mathematical product, theorems and proofs. One explained: *A group of people try to induct you or persuade new members that certain things are aesthetic.* Another mathematician said:

> *I think a reasonable proportion of people in the same field would agree about something that was beautiful. I think beauty probably requires a culture before you can appreciate it. Pictures and music are the same. I would have thought that no-one in mathematics would appreciate anything as beautiful that isn't really very simple.*

Together with such 'recognitions' come the behaviours of the mathematical culture which lead towards judgmental evaluations and, sometimes, consequent personal positionings. It is clear from what the mathematicians say that the judgments and the feelings associated with them are intimately connected: *Beauty is a reaction, you can't necessarily analyse it, you just feel it.*

In earlier chapters and in Chapter 6, discussing the epistemological model with which this study began, I have emphasised that it is neither useful nor straightforward to try to separate out the different categories although I have dealt with some of them separately. Neither do I believe that distinguishing the cognitive from the affective is helpful. In the case of the two model components, the aesthetic and the intuitive, they seemed inextricably interwoven, intuition providing, for many, the convincing energy that motivated and justified the necessary work and aesthetics producing what a number of mathematicians called the 'euphoria' (see below) that accompanied problem resolution. Consequently, although bound together, for most, but not all, of these mathematicians aesthetics and intuition appeared to fulfill different roles. The excitement of recognizing aesthetics was more often connected to the proof, the mathematical outcome. The intuitive leap, during the enquiry process, produced its own positive feelings. In Chapter 6, Sonia, like many of the mathematicians in the study, describes the cognitive processes of fitting, relating and connecting in terms of the affective expressed as joy or relief and for Sarah, intuition brings her the broad, connected and aesthetic understanding that, in turn, provided the motivation and pleasure of mathematics. The energy for the intuitions that sustain the search is the energy connected to what might be possible, whereas, when the aesthetics are attached to an outcome, they result from applying that energy and achieving a result.

INTUITION/INSIGHT

Many mathematicians have written about the importance of intuition. Reuben Hersh (1998), for example, called intuition "an essential part of mathematics" (p.61). For him:

> 1. Intuitive is the opposite of rigorous...
> 2. Intuitive means visual...
> 3. Intuitive means plausible, or convincing in the absence of proof...
> 4. Intuitive means incomplete...
> 5. Intuitive means based on a physical model or on some special examples. This is close to 'heuristic'.
> 6. Intuitive means holistic or integrative as opposed to detailed or analytic...
>
> In all these usages intuition is vague. It changes from one usage to another. One author takes pride in avoiding the 'merely' intuitive [...] Another takes pride in emphasizing the intuitive. (pp. 61-62)

Pointing to the above outcomes of applying intuition, he noted that they are much less specific about where its importance lies.

Rosemary Schmalz summarized a number of definitions of intuition in the following way:

> intuition is that faculty of the mind for which comprehension is spontaneous and immediate as opposed to rational and linear, and very often, though not always, sudden. (1988: 34)

And, in a note, the editor of the journal in which her paper appeared drew attention to the terms in which cognitive scientists such as Minsky, Schank and Lakoff "recast these ideas" (Ibid: Note 2, p.34). Rosemary Schmalz went on to discuss what she called "disciplines undergirding intuition" (Ibid: 35). To capture intuitions, she claimed that the first two of these disciplines were attention and commitment. She said "commitment makes possible the long stretches of attention, attention with a certain element of passion, which facilitates the intuition's functioning" (Ibid: 39). To these, she added detachment:

> it does not mean the absence of energetic commitment...working in a relaxed way without anxiety and not being so attached to prejudged results that one fails to be surprised. (Ibid: 39)

and its close relative, inner confidence that "allows one to leave a traditional path or attempt a new method or procedure" (Ibid: 41). I am uncomfortable with referring to such dispositions as attention, commitment, detachment and confidence as 'disciplines'. I believe this choice of word unhelpfully reinforces a notion of control and instruction that is consistent with pervading pedagogy in mathematics classrooms where working with or on student intuitions is rare. Also, discipline implies separation from feeling and being which might not have been Rosemary Schmalz' intention but is meaning carried by the word. I prefer to refer to dispositions and return to them below.

Raymond Wilder distinguished three roles for intuition. These were in helping to evolve concepts, in research and teaching, and in separating the 'knowledge component' from the 'intuitive component' of mathematics learning (1984: 41-45). He said that:

> without intuition, there is no creativity in mathematics [but] the intuitive component is dependent for its growth on the knowledge component. (1984: 43).

Douglas Hofstadter (1979:560) pointed to the need for a sense of 'judgment', 'simplicity' and 'beauty' and asked "Where do these intuitions come from?"

However, not all mathematicians refer to intuition. Some, like Ian Stewart (1995) who also concentrated on the mathematical outcome rather than the human process, talked of 'instinct':

> The entrepreneur's instinct is to exploit the natural world. The engineer's instinct is to change it. The scientist's instinct is to try to understand it – to work out what's really going on. The mathematician's instinct is to structure that process of understanding by seeking generalities that cut across the obvious sub-divisions. (p.13)

In what ways, if at all, did intuition or insight feature in the discourse of the mathematicians around their coming to know? Did the females and males differ in this respect? Again, as in other respects, heterogeneity of views was marked. In answering questions such as:

Is intuition important?
What is it?
Where does intuition come from?
How reliable is it?

the mathematicians offered a view of mathematics in the making which is, I believe, intellectually and personally more coherent and more holistic than the reified mathematical product presented in texts and to learners in classes. I use these questions to structure the mathematicians' responses. In most cases, either intuition, insight and, occasionally, instinct featured in their discourse when they were asked to talk about their work on a particular mathematics problem. But, occasionally no reference was made either explicitly or implicitly and, in these cases, I noted to the mathematician that they had not referred to intuition and wondered why. Where they made a reference either implicitly and/or descriptively, I asked: *Would you call that intuition?*

The use of intuition or insight as a category of the model that describes ways in which mathematicians come to know also permits a challenge to the positivist control of the epistemological agenda. Furthermore, historically mathematics has been understood as pre-eminently cognitive and, inevitably, this has been linked with masculinity. In this section, I challenge these connected sexist stereotypes, male/cognitive, female/intuitive. I believe this perspective helps to explain some of the labelling of mathematics as a male domain.

Is Intuition Important?

Fewer than 5 out of the 70 mathematicians took a completely negative stance saying something like: *There is no such thing as intuition in mathematics.* As with the other categories of the model, there was no apparent relationship between their gender and what they said about intuition. To emphasise this, of those rejecting intuition, a female mathematician said: *I don't think intuition plays a part* and a male observed: *I think intuition is a sort of word which is loaded, fashionable for people to disown or adopt.* It is difficult to be exact about numbers because of the exploratory nature of the conversations. It could be that a participant who asserted that intuition was irrelevant to mathematics still recognised something that others might call intuitive but labelled it quite differently and that I failed to discover this during our conversation. However, 65 of the 70 (83%) were comfortable with the notion that something operated in their learning process to which they were happy to give the name intuition, or insight, or to describe in ways similar to how those words were being used by others.

There were some mathematicians who did not deny the phenomenon but were concerned about the meaning carried by the use of the word 'intuition'. I began to wonder about the reasons for statements such as: *I don't think I would call it intuition*, or *Intuition is not a word I would use* or, in linking intuition to *something mystical*, rejected its use in a mathematical context. The resistance by some mathematicians to using the word 'intuition' to label something they accepted but preferred to call 'insight', or something else (see below), seemed to me possibly to constitute an intrusion of the social stereotype female/intuitive, male/cognitive perceived as inappropriate for a subject, mathematics, understood as supremely cognitive. If this is the case, it is a way of formulating the mathematics discourse that ensures the dominance of words, such as logical, rational, 'objective', associated with (male) power.

Many made statements that agreed with Reuben Hersh who said: "If we look at mathematical practice, the intuitive is everywhere" (1998: 61). The observations of three mathematicians provide examples:

> *One must use intuition all the time.*
>
> *The ability to pick up that kind of connection in mathematics is mathematical intuition and is a central feature.*
>
> *I don't think you would ever start anything without intuition.*

The only substantial work on intuition in mathematics education is to be found in Efraim Fischbein's (1987) book, *Intuition in Science and Mathematics: an Educational Approach.* He also noted similar contradictory perspectives:

> According to Poincaré, no genuine creative activity is possible in science and in mathematics without intuition, while for Hahn (1956) intuition is mainly a source of misconceptions and should be eliminated from a serious scientific endeavor. (p.4)

Despite this, amongst the participants in the study there was overwhelming agreement that intuition, or something that acted in a similar way but might be given a different label, was an important feature of the ways in which they worked on their mathematical problems.

What is it?

One mathematician explained: *seeing the best way forward – that's intuition.* Others spoke of 'gut feelings': *I may not have used the words intuition or insight but when I talked about gut feelings I meant what I understand people mean by those words.* Another attempted to explain such feelings:

> *There was a gut feeling that we had cracked the language in which we were expressing our concepts, so we knew we were right. The gut feeling comes from seeing that the notation has such parallels that the structures must be the same as now revealed in the notation.*

But, again, to emphasise the heterogeneity, I was told: *It is much more formal than a gut feeling.* Some mathematicians, in describing their intuitions, spoke of *Ahas!*

> *Mainly you get the Ahas! when you wake up in the morning. Think of the problem before I go to sleep, wake up with a solution which hasn't yet been incorrect. More often than not, I'm right*

Of the 70, 10 favoured the term 'insight', although they did not always describe it, but relied on metaphor. Making connections was important:

> *Insight is seing a connection.*

> *If the light switches on when I look at a problem, I have had an insight.*

> *Having an insight is having a feel for how things connect together. I think it is probably possible to manage without it in that, by sheer bulk of discovery and trial and error, the same picture might emerge but that flogging to death involves a lot more work, a lot more blind alleys. So the insight is a short cut.*

Vera John-Steiner (1997) suggested that: "Metaphors, analogies, and the delineating of new patterns is the stuff of insights" (p.203).

There were those mathematicians for whom insight and intuition were distinct:

> *view, which is a linking of insights. I would use the word 'intuition'. I think the two are connected. You are always thinking in a not-straightforward, deductive way. You are always looking for some hint from within that this might be an interesting thing to do, or you should check up on this, that sort of thing.*

> *Insight is a sort of flash; intuition is more murky and definitely develops because if you have a false one, that intuition is modified.*

Others preferred to avoid using the words intuition or insight, sometimes mentioning the *'mystical connection'* or, in some other way, making their discomfort with the words clear. One referred to *I wonder ifs,* and they ruminated on how to describe this process which they recognised:

> *Ponderings maybe.*

> *An enhanced understanding that perhaps comes to you suddenly.*

Pattern matching is the best way of describing it.

Most, while recognising something that they called intuition and/or insight or one of the other variants above, did not agree on whether they were distinctive or even different nor did they share descriptions. However, some words, such as *'connections', 'ways forward', 'understanding', 'a sense of the possible or even likely',* were shared. While agreeing that intuition was not always sufficient, they found their intuitions (insights) reinforcing in that, through them, they achieved a sense of direction. But one warned: *intuitions can come close to prejudice, which can be a very blinding thing.*

Where does intuition come from?

Reuben Hersh offered a cognitive explanation:

> Intuition isn't direct perception of something external. It's the effect in the mind/brain of manipulating concrete objects – at a later stage, of making marks on paper, and still later, manipulating mental images. This experience leaves a trace, an effect, in your mind/brain. (1998: 65)

One of the participants agreed with him:

> *I know the sort of things that will and won't work so my instincts are probably better than they were. Instinct is not the right word. Insight is better. On the basis of what I know, and what I have done before, I feel pretty sure that this kind of method will work on this kind of problem. It is not intuition because it comes from experience. There is nothing mystical about it. I have seen what works and what doesn't. It mustn't carry any suggestion that there is some otherworldly process going on. There are things that go on in the brain that can't be clearly explained but they are processes which have been polished by past experience.*

For most, however, the combination of knowledge and experience was exactly what *did* explain their intuitions: *You might have an intuitive feeling but it would be based on your knowledge and your experience.*

> *My intuition is a result of a synthesis of previous results that I have shown or putting it together with what other people have shown and applying it to this particular case.*

> *You have ideas and then a new idea helps you to make greater sense of the ideas you are currently holding. So it is some kind of synthesis between your experience and your knowledge.*

> *What makes me a better mathematician now is that I have solved a lot of problems, worked in different areas, so now I have a lot of experience, I have intuition about what is going to work. I don't think it is as mysterious as it sounds. You develop it by just doing it. If I look at my older colleagues, what they have is that experience. They look at something and see the best way forward. That's intuition.*

So a combination of knowledge and experience was how most participants accounted for their success in achieving insights or intuitions. A few mathematicians believed that you either did, or did not, have 'it', which was presumably a genetic explanation. Some of the female mathematicians recognised intuition in others. For themselves, they spoke of: *Gut feeling - I feel all of these things come from experience. I might pin mathematical intuition or insight on other people but not on myself.*

How reliable is intuition?

According to one mathematician: *The two, intuition and predictability, go hand in hand.* Or, as Brian Rotman put it: "mathematical assertions will be understood as claims about the future" (1993: 77). While many of the mathematicians adopted the position that intuitions were necessary to doing mathematics, none thought that they were sufficient, particularly because of their unreliability but, of course, also because of the mathematical cultural conventions which dictate the format of publication of results.

> *I think one has intuition about how the physical world behaves. I certainly do. You acquire it very slowly usually because you have done a number of calculations about related systems and you discover that something ought to be true because of that. It is often wrong and NEVER is a substitute for a proof.*

> *I have intuitions but they are not always reliable. They require a lot of checking. My intuitions about differential geometry are as likely to be true as not. Maybe a bit better on Partial Differential Equation theory. My intuitions are based on my knowledge and my experience. The more I have, the more robust my intuitions are likely to be.*

> *Sometimes I know that something is true even though I haven't proved it. I suppose that would be called intuition. It probably comes from experience. The more experience you have in an area, the more likely you will make correct conjectures. So I can believe that something is true without having proved it. Sometimes I am wrong in this or I cannot prove it. Sometimes it is correct, sometimes I manage to prove it. Experience is probably having*

to do with knowing something similar and then you can guess probably that the conjecture is true. That is about making connections.

Whether or not you can trust your intuitions was something that featured quite prominently in our conversations but more especially for the females. There follow three statements, all by women:

You might have an intuitive feeling, but it would be based on your knowledge and your experience. But I think you need a firmer basis for doing something than intuition but perhaps I am too cautious.

Sometimes intuition is wrong. You think something is going to work one way and you battle on and it doesn't work. When I do that, trying to see how it really is and trying to explore, I think I am trying to get an insight into what is going on. One must use intuition all the time. My intuition would be something that says I think the argument might work this way.

I don't think I am intuitive enough. If you know something is a good problem, or not, you are intuitive – have a good sense of whether something can be done. I don't trust my own intuition [...]There are intuitive mathematicians.

Nonetheless, there was widespread agreement that having good intuitions was a function of knowledge and experience both of which are the product of learning. This leads to a growing sense that the pedagogic contradiction that, on the one hand, asserts the importance of intuitions to making mathematical meaning *and* places knowledge and experience at the heart of the making of intuitions but, on the other, ignores intuitions as an area for learning, needs urgent attention. (I develop this in Chapter 10.)

HOW DOES IT FEEL – TO BE A MATHEMATICIAN?

Maybe an intuition comes from inside yourself, no external stimulus. Intuition is to do with feeling, you have a feeling that something makes sense but you cannot quite explain why. I think I prefer to use the word insight whereby we connect what we know previously with what is happening now. Connecting what you know with current knowledge in a way that brings greater understanding.

I began this study with a conjecture that the process of doing mathematics was a very human one and, consequently, affected by the concerns and feelings of those involved. The epistemological model that the study was designed to examine from the perspectives of the mathematicians was not a cognitively based model but rather one that drew on factors which positioned, and were positioned by, the mathematical researchers. It focused upon how they come to know, not on their acquired knowledge. Mathematical activity, for many of these mathematicians, was motivated by curiosity and the resultant pleasure when something was resolved or strong feelings of frustration when an intuition proved untrustworthy. Contrary to what many teachers say, frustration was not to be avoided but was a natural part of the enquiry process. They described in very emotive terms their satisfaction with finding a pattern, making a connection, eradicating a difficulty, in general with coming to know:

> *There are lots of different ways in which one understands something. The most gratifying is that sudden wave of insight in which suddenly something all becomes clear. That is sadly rather rare.*

As they made clear, coming to know invoked their feelings, and involved their perspectives, not only in what they chose to research, how they did that research, how they reported on their research, but most of all how they felt about it. In discussing research focus, one mathematician said:

> *... not helping practitioners is important to me. The more the technology advances, the more people become unhappy. I would not like to produce something which could be exploited or misused by people over whose actions I have no control.*

So what the mathematicians said about their practices in mathematics appeared to be in contradiction with the widely presented and believed image of a cognitively dominated and "objectively" clean discipline. Indeed they bore out that

> mathematics is not so much a hard crystalline mountain of truth painstakingly built up by successive generalizations as a field for error, for the conflict of ideas, for vagueness, for continual approximation and conjecture as the means of progress. So it's recognized among mathematicians, if not outside, that we do mathematics one way, but we might well do it another. (Hodgkin, 1976: 30)

The affective and the cognitive were indivisible for them. Far from understanding being something that is *only* driven by knowledge, there is both a *need* to know and an associated *pleasure* in knowing which is its own reward.

> Most mathematicians do mathematics for the very good reason that they like and enjoy doing it. (Mordell, 1984: 157)

For many, the intuitive or insightful leap appeared to be integral to this, even though it was not always reliable and also might be disconnected from proof or from the need to justify through rigorous argument. Indeed, George Polya offered a quote from Hadamard to the 2nd International Congress on Mathematics Education[2] that made the pursuance of "the conquests of intuition", rather than the production of a rigorous proof, the foremost mathematical activity:

> The object of mathematical rigour is to sanction and legitimate the conquests of intuition, and there never was any other object for it. (quoted in Howson, 1973: 78)

In these circumstances, the question of how to feed intuition and nurture the resultant enquiries, not only how to produce convincing proofs, is a central one, both for research *and* for teaching. I return to this in Chapter 10.

Mathematics as a discipline is both responsive to a humane, socio-cultural perspective balanced in its acknowledgement of the role of people as well as ideas, and contains the products of the work of those people, the body of knowledge revered by mathematicians as absolute. But: "If mathematics has to be spoken of in the vocabulary of 'truth', then this can only be truth as a human artifact, truth as a description and production of the one who proclaims it" (Rotman, 1993: 158). Acknowledging that mathematics has human connections demands an acknowledgement of both the messiness of human feelings and that the mathematics itself is not unconnected from the effects of the emotions. The importance of this to all, mathematicians and non-mathematicians alike, remains largely unacknowledged perhaps because it opens up mathematics to admit the play of emotions alongside cognitions. It is a fundamental challenge to the subject-image built upon objectivity and abstraction and has implications that are explored in Chapter 10 for how the subject is understood and presented, for how it is taught and learned.

Many of the mathematicians used extremely emotive language in speaking of their discipline and their practices within it. For example, with respect to reviewing a paper submitted for publication, a mathematician said:

> *It would be a beautiful argument. It is crystal clear with clear lines and looks beautiful, appeals aesthetically but it would have these properties independently. Light and clear, perfectly balanced and poised and would slip into your mind and you would be very pleased in the same way as you react to nice music.*

---

[2] I am indebted to Bob Burn for reminding me of this.

It is difficult, here, to dissociate the description of *'argument'*, a necessary component of mathematical proof, from the feelings driving the choice of words. Indeed, I would claim that such dissociation would be a distortion. Of course, when she writes out a proof and makes an argument in formal terms, the community, and consequently she herself, places certain demands and expectations on how this is presented. However, in the statement above lies the sense of what drives her interests in mathematics, and what it is that she aims to produce. Another participant said: *One of my tutors took me for topology and we worked very well together. I enjoyed the topology but it was also that we were relaxed and enjoyed it together.* I believe she was making clear how much the mathematics that she learnt was inter-connected with the associated feelings invoked. To deny such feelings is to turn mathematics into the cold, inert subject described by so many people who have attempted to learn it (see, for example, Buerk, 1985).

My claim throughout this book is that the feelings expressed in such statements are a learning necessity. To try to learn mathematics without feeling any passion is a contradiction. The mathematicians have confirmed this sometimes very explicitly. In examining their feelings below, I draw on the sparks expressed as excitement, satisfaction, and euphoria that featured in their positive discourse. Such feelings are frequently associated, as has been seen, with aesthetics and/or intuition.

Excitement

> *I find mathematics exciting. It would be dull to wake up and not be thinking about this world I call mathematics.*

A sense of excitement was something that many of the mathematicians invoked when they were talking about their research. This sense could be located in doing something original, feeling special, sharing something with others in the field, a motivator to keep you on track when things are not going so well and a reward for the achievement. One of the exciting things seemed to be the making of connections.

> *Something clicked in my mind that this was actually a pattern, two infinite families and three exceptions. I was especially pleased because I was the catalyst - I saw the link. It is the pleasure of seeing something like that. Seeing a connection is really no different from proving a theorem.*

Linking, as this does, with the model category of connectivities, it is both reassuring and reinforcing of the model.

Interestingly, it is very rare to find students speaking in terms of excitement but they sometimes do indicate that the rewards that come from achieving understanding might be similar. For example, Jo Boaler reported Philip, a 16 year-old, saying:

> It's when you like learn new ways of doing things or you're like doing quite well on a problem...you're taking it really far, the investigation, you're getting really deep into it...you feel like you're learning quite a lot more. (1997a: 58)

There is not just a sense of satisfaction to be found in that quotation but also a sense of possibility, the kind of possibility that, for the mathematicians, leads them towards excitement and the importance of sharing:

> *When it works, it feels great. It is very exciting. It doesn't happen that often but when it does it makes it all worthwhile. You write something down that you are pretty sure has never been written down before, it is very exciting, and the excitement is shared by other people as well.*

> *If you have been excited about doing something, it is good to let off steam sharing with colleagues or my wife.*

Unfortunately, that sense of excitement tends to be discounted as a necessary part of the process of finding out or, at the least, to disappear in the public presentation of mathematics. So excitement often remains very much a part of the private domain within which the mathematician works but upon which comment rarely surfaces publicly. That this is an unsatisfactory aspect of conventional learning, the next participant makes clear:

> *It is quite different from learning as an undergraduate [..]. You are a bit privileged with some piece of knowledge. It's nice to know that you can tell other people about that instead of them telling you about their work. I suppose I felt a bit clever and sort of excited for myself and I thought that the other people I was working with at the time would be excited.*

Satisfaction

As we have already seen, satisfaction tends to feed excitement, the one often leading to the other and, possibly, the other way around.

> *The reward would be totally self-contained, the satisfaction in having completed it, of knowing that it works.*

Many of the mathematicians spoke of satisfaction and it is possible to trace a number of factors that create this feeling. There is:

> *Consistency because when you meander through the complex plane there are certain places that you know you must get to and as long as it gets to all those places you know that you've got a route through to the other side if you like - the joy of actually creating something - of building a structure, of completion.*

In addition, there is that sense of structure:

> *I suppose one gets a sense of exhilaration, an appreciation of its elegance - a fit of form and shape in a very abstract sense just as in certain paintings or music you have got a structure which holds together with a sense of both differentiation and balance and wholeness.*

But the beauty and balance of the structure is not the only reason for the way it contributes to a sense of satisfaction. In addition, there is a growing sense that *It is partly a matter of form - if the form is not correct, you can see straight away that there is a mistake... that was a real joy finding a whole new structure that hadn't been found before.* Most of all, mathematicians spoke about completion:

> *When it actually comes, it has a sort of feeling of completion and correctness.*

> *There is certainly satisfaction when one gets a result and knows it is a result.*

> *There is a feeling of general pleasure in completing a nice piece of work. It is an achievement. Something that hasn't been done before.*

> *When you complete a piece of work, you feel 'I did that' - it is very satisfying.*

So, satisfaction is rooted in a sense of consistency, structure and completion that also, it will be recalled, provided aesthetic functions. Satisfaction, for some, is the positive feeling that arises as a result of reaching a conclusion: *it's like climbing a mountain - it's hard work but it is exciting, rewarding, enjoyable. The reward comes at the end.* But one mathematician warned: *Achieving is a reward, but it is doing mathematics, rather than being smug about having done it, is the experience that I enjoy.* But a conclusion does not have to be a final product. It could be the sense that the progress that is being made is leading in the right direction. And although many of the mathematicians, like those below speaking of euphoria, were brutally honest about how many times this feeling let them down, nonetheless:

In public, I would only say I know once I have written down the proof and I am absolutely sure. But there is a stage before that when I would say it is going to go but it is merely a case of writing down the details. You haven't got a written proof yet but based on experience as much as anything you can say how the written proof could go and perhaps the big obstacle you are trying to surmount you can see perhaps one small detail that completes the proof. It is a very good feeling; I wouldn't do maths if it weren't for that.

Euphoria

I found it interesting that more than one mathematician used the word 'euphoria' to describe the poignancy of their feelings when they think they have achieved some knowing. Not many members of the lay public, nor students in schools, would think that mathematics is a context for euphoria and yet this is what was said:

When I think I know, I feel quite euphoric. So I go out and enjoy the happiness. Without going back and thinking about whether it was right or not, but enjoy the happiness.

When I discover something, I just enjoy the feeling. It is almost pointless to try and check it because you are so euphoric that you cannot possibly check.

When I think I have solved a problem I go completely euphoric. Immediately, I have to communicate this excitement. Sometimes, I am wrong and have to start again.

Speaking of the benefits of collaboration, another mathematician said: *Partly it is someone to share the euphoria with.*

AFFECT AND MATHEMATICS

It seems appropriate here to re-introduce the four dispositions, outlined by Rosemary Schmalz and mentioned above, as contributing to the capture of intuitions. They were attention, commitment, detachment and confidence. These have already featured in many of the quotes from the mathematicians and help to tie together the three sections of this chapter, on aesthetics, intuition and feelings. A combination of all four dispositions explains what the following participant described as he talked about the intuition, the *kind of a hunch* that first sparked a collaborative effort:

> *We were both absolutely convinced that this did something spectacular and it did and we knew it from our first conversation. We spent the next two or three weeks working quite hard on this and we had to work out the details but both of us were aware of the intricacies of the mathematics. One conversation was enough to convince both of us that we had something that we were working on and it would go somewhere. If you had asked us why we said that, it would have been based on - well we had these ideas and we need to work this out. We didn't ever question that this would work. I am not sure how much evidence we really had. It was a kind of a hunch and we could talk sensibly to each other.*

The feelings associated with the result of applying attention, commitment, detachment and confidence also feature in the next example:

> *At a gut level, I am more convinced by knowing lots of examples that illustrate the theorem I am trying to prove even though they don't prove the theorem. Just because there are lots of examples that show this as being true, doesn't prove it is true, but it makes me much more comfortable. So if one is very familiar with the space of likely examples and their extremes, then I feel much more confident that the result is true.*

For these mathematicians, and many others, intuition embodied the subjective and through having reliable intuitions, the subjective was transformed into the production of 'objective' mathematics. Reuben Hersh explained this by saying:

> *The study of mental objects with reproducible properties is called mathematics.* Intuition is the faculty by which we consider or examine these internal, mental objects. (1998: 66, original emphasis)

But the demand for infallibility and objectivity in mathematics requires the removal of the evidence of the very human beings who created and worked with the mathematical objects, missing "entirely the stumbling human process that created those results in the first place" (Smith & Hungwe, 1998: 46). Furthermore, while they might acknowledge the 'stumbling' for themselves, none of the mathematicians spoke about a need to work on their, or their students, intuitions either to improve their frequency or their reliability although one mathematician did refer (above) to an intuition that failed as providing the potentiality for improvement of that intuition. The vagueness of the following explanation perhaps gives a sense of why there is so little acknowledgement of the pedagogic contradiction mentioned above:

If you are doing a piece of research your intuition can tell you whether you are going up a blind alley or not. Although you can't work out the details, you realise that you are about to hit a dead end and you will have some hand waving reason why but you know in your gut that it isn't going to work and equally you know in your gut when it is going to work. You might not be able to pin down the details yourself (that is often when collaboration comes in because they might be good at that part of it) but you know somehow that that is the right way to do it.

From the three positive themes, excitement, satisfaction and euphoria, expressed sometimes as an individual reaction and sometimes in the context of the importance of collaborative practices – someone with whom to share – we get a strong sense of the feelings and the results being inter-connected, one dependent on the other. Inevitably for some, there were negative feelings as well and these are explored in Chapters 7 and 9. The irony that these positive connections between feeling and thinking are unavailable to learners is addressed in Chapter 10 along with the positive implications for learning of the many strategies and practices of the mathematicians that provided bridges between their aesthetics, their intuitions, the feelings these provoked and the learning that resulted.

SUMMARY

In this chapter, the affective context within which the mathematicians worked was connected to their development of knowing through an exploration of two of the categories of the epistemological model, aesthetics and intuition. Most of the mathematicians rated highly the importance of both aesthetics and intuition and they freely explained how both evoked feelings which then fed, and were fed by their attention, commitment, detachment and confidence. These, in turn, encouraged and substantiated their intuitions. Aesthetics were discussed from four different perspectives, generative, motivational, evaluative and functional but the last was the one most frequently invoked. Functional aesthetics were described in terms of succinctness, generality, structure, connections and simplicity all of which are 'learnt' as part of the culture of mathematics. The links between the aesthetic perspectives and the dispositions of intuition were reinforced by the inter-relationships demonstrated in what the mathematicians said. The mathematicians affirmed the importance of intuition to their mathematical thinking although some expressed their preference for the word 'insight' over 'intuition'; I conjectured that this could be a product of sexual stereotyping. They affirmed that intuitions, or insights, were a product of knowledge together with experience and were not always reliable. We saw again how the intensity of their feelings interacted with their learning. Their positive feelings were expressed in terms of excitement, satisfaction and euphoria.

Many implications for mathematics education arise from this chapter, particularly in the strategies that the mathematicians described as producing beautiful or elegant mathematics, as provoking their intuitions and/or as contributing to their positive feelings. They mentioned, for example, structures that help them to build a convincing argument and the importance of asking questions. Again, they reinforced the importance of connectivities, the model category on the importance of which they all agreed. The issues raised by the mathematicians in this chapter and the emotive ways in which they raised these issues guide us to a pedagogic perspective together with a critique of the mathematical experience as it is offered to many, if not most, learners. In Chapter 10, I discuss the importance of creating a culture in mathematics classrooms that is both intellectually *and* emotionally rewarding and consequently more productive for learning.

In Chapter 6, through case studies of 17 of the mathematicians, we meet those whose personal data help to provide a context for each of the model categories, and then, holistically, for three mathematicians.

6

Meet the mathematicians

written jointly with Nathalie Sinclair

We felt that it would be informative, through case studies, to bring the mathematicians as individuals into these pages. We wanted their voices to be heard much more directly than in the other chapters where comparative interpretation is involved. We see what they have to say both specifically, with respect to each model category and then globally, that is across all five categories of the model. The intention was not to establish characterisations or general trends within each category so much as to communicate some of the richness and complexity of each category, for each of the mathematicians, as they describe how they come to know throughout the interviews. Also, meeting the model in this way, through a personal approach to individuals, we hope helps to bring more meaning to the categories and to the model as a whole.

We begin by presenting brief case studies of mathematicians who provide more insight into the ways these mathematicians articulate each aspect of the model. This does not mean that they did not discuss all five categories as, indeed, most of them did and even within the presented case studies sometimes other model categories appear; but we have chosen those portions of the interviews which offer good examples of each category of the model in turn. To exemplify the inter-connections, the chapter concludes with 3 case studies across the model as a whole. Sometimes a quote that has appeared in earlier chapters will be heard again because it was said by the case study mathematician and is particularly relevant to the ways in which they present themselves.

PERSON- AND CULTURAL/SOCIAL-RELATEDNESS

We introduce Annette and Edward, two mathematicians who discuss the first category in the epistemological model, its person- and social/cultural relatedness. Unlike the prevailing view of mathematics as absolute, objective, "a body of established knowledge accessible only to a few extraordinary individuals" (Rosser, 1990 quoted in Burton, 1995, p.276), mathematicians who recognise the relationship between the discipline, themselves and the culture and society within which they are placed are using this category to understand knowing as a "function of who is claiming to know, related to which community" (Ibid, p.287).

Case study 1 – Annette[1]

Annette was not born in the UK – coming to England only to take up a fellowship that permitted her to begin a Ph.D. She had wanted to study mathematics from *quite an early age* and had achieved well at school. Her family was not academic *so there was no-one to advise.* She *did as much discrete maths* as she could and ended up doing a {Master's} thesis on Design Theory. Encouraged by one of her lecturers, she registered for a Ph.D. in England. Unfortunately, the funding agency withdrew her fellowship after one year and she had to take employment. She claims:

> *I became a mathematician without choosing. It was more like that is what happened. I don't like all kinds of mathematics and I don't understand how mathematicians can like that stuff. I think I am a combinatorialist, not a mathematician. What I like about combinatorics is that I see it a bit like solving a puzzle. There is a challenge. The problems are easy to explain to a lay-person but trying to solve them you need a lot of various techniques. It is a bit like having a mystery that everyone can understand, but they don't know who did it!*

This introductory sketch of Annette's approach to mathematics makes clear that it was highly person- and socio-related. She wants to be able to talk with others about her problems but *it could be anybody; it doesn't have to be somebody knowledgeable in the area.* She says that she has *always had people to talk to about my work so it never feels as if I am on my own.*

Annette appears to have had a fairly conventional socialisation into mathematics and, like many of the mathematicians, she explains *the pleasure of doing mathematics as something like getting the last piece in a jigsaw puzzle.*

---

[1] All names are fictitious.

(See Burton, 1999b) She compares the enjoyment with that gained from reading mystery novels. *It is a game, you bring some clarity to a foggy picture. There is an aesthetic quality as well. She used to think that mathematics was the only certain science* but had a Damascus-style revelation as a result of doing a course on the philosophy of science where she learnt about Gödel. She says:

> *I think it is very common for people not to question mathematics; what is there is true absolutely. When you get further in, we find that everybody is standing on the top of everybody else's shoulders, and there is no uncertainty. Then the bombshell falls.*

As a result:

> *it doesn't now matter to me that it isn't certain. It is the challenge of building the tower, not whether or not the tower will fall. You have theorems, which you can neither prove nor disprove. You have one in combinatorics. When I saw one in my own subject, you get down to something which is very difficult to grasp.*

Because of her need to re-think her relationship with mathematics, Annette now adopts a very person-oriented approach to mathematics. She says:

> *I think that I always thought that maths was very beautiful. The process of deducing one thing from another, so before I knew about Gödel, I probably thought it was the most beautiful thing in the world. Now, the gilt is off the gingerbread but, nevertheless, it is very satisfying to know that I have something that I can tell you about and, if you will listen to me, I should be able to convince you. You should be able to understand, if you accept my foundation. Mathematics is like that. Life is not like that.*

Case study 2 – Edward

Edward, too, was born outside of the UK, coming to England as a post-doc. and taking up a number of temporary lectureships before finding a tenured post. His mathematical history before arriving in England was gained across two continents and this, possibly, contributed to his very original way of describing his understanding of mathematics. His area was groups and geometry – *the purity isn't so important. I became more taxonomic.* He says:

> *My images of mathematics are my experiences of mathematics. So they really do change. The kind of mathematics I do now is certainly different from what interested me previously. My mathematical world has changed – it has become broader as you read more. You can happily do research but*

> *not learn things. Some of the core concepts, things that I started hearing about in my first year of the Ph.D., have stayed with me and are becoming more sophisticated as I go along. You do realise that, in fact, you did not previously understand as much as you thought you did, and there was more in it than met the eye.*

From statements such as these, Edward shows himself as someone whose images of mathematics are not certain and absolute. *It is a case of puzzling through, asking yourself why. How do you know that a new generality is more robust? You don't really.* This is a mathematics of the person, fallible and able to be influenced.

Edward prefers to *talk stories* and likes *articulating things in words rather than formulaically.* He does not want *to be perceived as a loner.* He is *very conscious of how the ideas are expressed. I like a story – a paper, a seminar are both stories. There is a beginning, a middle and an end.* Edward gives the strong impression, like Annette, of recognising how important it is to be able to communicate your mathematical thoughts, and your mathematics, to others. But he also cites connectivities as the grounds for something being worthwhile to do in mathematics. *The satisfaction achieved in finishing the work is enough for finishing but I want to know how it connects to other things.* This, again, is a socio-cultural motivation for engaging with mathematics.

Edward describes mathematics as *a subjective discipline or process.* He cites the role of powerful mathematicians in controlling who is heard and what is acknowledged as important.

> *There are certain areas, and mathematicians, who are considered to be better than others and often a piece of mathematics will be seen to be better because of the association. The hierarchical aspect is there all the time. If somebody said to me is mathematics objective or subjective, I would say subjective...but the issue of fashion and fad and reputation by association, prestige – there is a game which is being played and you make it hard for yourself by not playing it.*

Summing up: Unlike the majority of the mathematicians interviewed, both Annette and Edward locate their mathematics in people, not in objects and acknowledge the influence of the social and the cultural on their mathematics and on the mathematicians with whom they work. They also appear to be aware of the influence that such a location has on the social experiences of mathematics, Edward in particular speaking about hierarchy and competition. Many of the other mathematicians made spontaneous reference to social or cultural influences on their mathematics without, necessarily, being reflective about the implications of this.

THE AESTHETICS OF MATHEMATICAL THINKING INVOKED

Aesthetics is a term that can mean quite different things to different people. So, we cannot avoid the personal, social and cultural. Aesthetics is sometimes used as a catch-all phrase for referring to the harmonious and the beautiful, yet it can also have quite specific (even contradictory) philosophical shadings. Within that, there is the essentialist position that portrays the aesthetic as belonging to or existing in the mathematical object; a strong tradition in the mathematics literature supports this view (for example, Hardy, 1940 and Penrose, 1974). A contrasting subjectivist position sees the aesthetic existing in the perceiver of the mathematical object (for example, Le Lionnais, 1971). A third contextualist position sees the aesthetic as part of the particular historical, social, or cultural context (for example, see Weschler, 1978).

In Chapter 5, we gained the mathematicians' perspectives on aesthetics structured through a use of Nathalie Sinclair's (2002) tripartite model of the characteristics of mathematicians' construction of aesthetics. The three characteristics were the generative, the evaluative, and the motivational; the generative focused on the role of aesthetics in inventing or discovering mathematics; the evaluative typically manifested itself in the judgement of a mathematical product, such as a theorem; and the motivational related to the role of aesthetics in prompting or inspiring mathematical activity. Function, was also introduced in order to ask how does an aesthetic operate as a way of knowing? To exemplify the mathematicians' responses to such issues, we introduce Maria, Sonia and Graham.

Case Study 3 - Maria

Maria is a pure mathematician, a numerical probabilist who feels she drifted into mathematics and possibly, as she remarks, even *accidentally* ended up as a numerical probabilist. She studied philosophy at university, taught Primary 1 mathematics, and then took time off to be a mother before starting her work. Though surely she is a mathematician in the eyes of many, involved in both lecturing and doing research, she resists that appellation: *But I have met people whom I consider to be real mathematicians and I am not one of those people.*

Maria believes that mathematicians can be divided into two groups according to the way they think: one group is pictorial and the other group is language-bound. Maria puts herself into the second group, saying: *I think I talk. Most often I end up finding what I am looking for myself by hearing myself talk.*

When Maria is forced to use symbols, as when she works with the computer software MAPLE, she says that *those symbols probably become the words*, re-emphasising an essentially linguistic construction of mathematical meaning-making. She also finds that there is a certain amount of sub-conscious mathematics that goes on in her mind. In describing how she comes to know, Maria says that she translates results into her own terms that are not as rigorous. Although this might seem like an intuitive way of coming to know, Maria maintains that it is just more concrete: *a translation from the abstract,* a personalisation of the linguistic idiom.

Maria introduces the evaluative characteristic of aesthetics, although not using those words, when describing criteria she would use in reviewing a paper. She asks herself whether she would find the paper useful, not necessarily in an applications sense, but whether there is a certain amount of new material. But she cautions that what is interesting can also be faddish: she says that finance has become *the big thing* in her community but that she finds this area boring and does not intend to start publishing finance papers. She believes that mathematics is a product of people and that even though it does turn out to be extremely useful, *utility is also a cultural product.* She is suspicious of the word beautiful and says, *words like beauty and elegance are over-used.* A few sentences later she says that she is a very logical person and when she knows something, *it is just logic*. Knowing for Maria is mainly about prior learning and experience and logically making connections. She dismisses intuition as a way of knowing.

Maria says that she used to enjoy doing mathematics for *aesthetic reasons* as a child but that now, if she were to explain her enjoyment in mathematics, she would do it in terms of finding it relaxing and peaceful, and of the sense of achievement. She says that she does not think of mathematics as being beautiful in the same way that she finds music or pictures beautiful. Although as a child, Maria used to derive aesthetic pleasure in doing mathematics, it would seem that, now, Maria's conception of aesthetics is disjoint from anything mathematical in her work.

In her interview, Maria at no time indicates a use for any characteristics of aesthetics such as those described by mathematicians in the literature; she explicitly negates an evaluative characteristic of aesthetics in describing the criteria she uses to review papers. Having said that she *accidentally* fell into her subject area, she chooses problems according to *what is left to be done* rather than what might interest or appeal to her. Choice, in field of interest and in particular problems, is the primary locus in which other mathematicians have identified the motivational characteristic of aesthetics (Penrose, 1974; Von Neumann, 1956). However, in Maria's case, the motivational characteristic did not seem to influence her choice. A third characteristic, the generative, is often linked in the literature with intuitive or subconscious thinking processes and feelings (Poincaré, 1908/1956; Papert, 1978), and as a mode of cognition which focuses on forms, structures and metaphors used in conceptualising and modelling. Early in the interview, Maria does refer to *subconscious mathematics that goes on* in her mind but makes no reference to attending to the forms or structures of mathematics. Her primary way of knowing in mathematics seems very literal, attending, as she claims to do, to the outer skin of the forms and structures through language and symbols.

Buying in to the hierarchising that goes on in the community of mathematics, Maria hesitates about whether or not she is a *real* mathematician although this might indicate that she lacks a certain amount of confidence. She does not see herself as an *ideas person* but rather as a person who logically works things out. However, this logical, literal and persevering interaction with mathematics has perhaps been successful enough for Maria to encourage similar interactions and discourage other more intuitive or aesthetic ways of knowing. Or, she is a product of that particular approach which possibly dominated her learning experiences at an important time. If she has had no experiences with an aesthetic way of knowing, she might either distrust her own potential or even believe herself incapable of such a way of knowing. It is also possible that Maria's formal language-bound and concrete thinking style does not lend itself well to aesthetic ways of knowing. Although there is nothing particularly non-aesthetic about these thinking styles, she describes them in a manner that seems to preclude a more holistic and figurative way of perceiving mathematical relations.

Case Study 4 - Sonia

Sonia is also a pure mathematician, a graph theorist who enjoyed mathematics from an early age and encountered positive experiences both at home with her mother and at university with her mathematics tutor. She remembers meeting cyclic quadrilaterals in school at the age of 14 and realising that *circles had these wonderful properties and that she was delighted with the art of geometric proof.* Sonia started in topology, then moved to combinatorics and then finally into graph theory. She speaks of being obsessed by and in love with each of these areas, noting that she follows particular areas of interest somewhat accidentally but also influenced by personal preference.

Sonia works in a very collaborative manner, in part because she found her own experience in undergraduate and postgraduate work very isolating. She enjoys working collaboratively for many reasons: to have someone with whom *to share the euphoria and to spark ideas.* Sonia tends to think in pictures.

She acknowledges that there is a certain amount of symbolism that goes with the pictures but that she can also visualise symbolic expressions. She believes that mathematics is the abstract description of the pictures—framing a linguistic surface around a visual/spatial/pictorial heart.

When describing a recent problem on which she worked, Sonia says that the result was quite complicated and that she is worried whether others will find it interesting. Having solved the problem *tortuously*, she said that the next problem is to solve it elegantly, and that elegance is very important. Sonia thinks that intuition plays a part when one starts looking at a problem, and that indeed, she would never start anything without intuition. She says that she enjoys the original sorting out of the problem. Then she speaks of the combination of intuition and growing experience as one continues working with the problem. When she is stuck on a problem, Sonia goes away until a new thought or angle *strikes* her. Sonia describes her experience, at the young age of 8 or 9, of writing her *first theorem*; it was at that point that she realised that mathematics did not consist *of endless long multiplications and divisions of Imperial units of money and measure* but could be created.

From the beginning of the interview, Sonia talks about enjoying the artistic part of mathematics and about the excitement she gets from the beauty. She clarifies that beauty is not in the diagrams that look pretty on the board but in the structure connecting the beginning to the end, a mixture of the generational and the evaluative. She compares appreciating this structure with looking at something structurally beautiful such as seeing a beautiful picture or hearing a particular piece of music. She goes on to describe the aesthetic appeal of the chain of logic and the completeness she finds in her mathematical work. She adds that she chose her branch of mathematics because it had to appeal to her in an aesthetic way.

Sonia alludes to the evaluative characteristic of aesthetics in her statement that a problem should be proved elegantly. It would be helpful to have a better idea of what she means by elegance; from her description of the current result as being complicated and a *tortuous* argument, connoting pain and oppression perhaps, relying on a series of steps, we can infer that she prefers to have a more simple, liberating, and revelatory argument. She adds that elegance is *all very relative of course*. This would suggest that it is not the result itself that is elegant, but the way in which Sonia understands the result; she assumes a very subjectivist position as to the location of the aesthetic. For example, she might come to know the result as she is able to grasp its elegance, that is its lucidity, painless coming together, and liberating transparency. How is this way of knowing different from others? It is certainly not the same as the oft mentioned "I know it if I can prove it" way of knowing which is characterised by deductive reasoning, perhaps through logical transformations, based more or less on foundational assumptions. But how is it distinguishable from an intuitive way of knowing? It is apparent that trying to separate the aesthetic and the intuitive, or indeed the other categories of the model, is not easy nor always useful; they seem to operate in tandem. However, it is helpful, to understand the function of the aesthetic, to tease apart possible distinctions. According to many of the research participants, intuition seems to provide the conviction that something is true or provable, enough to motivate and justify arduous work. It also often operates at a covert level. What Sonia is describing as aesthetic has two distinguishing features: there is a definite affective (motivational) component that manifests itself as joy, relief, and integrity; and, there is an active cognitive (generative) component of fitting, relating, and connecting.

Sonia's descriptions of how she chooses problems and areas of research match well with the motivational characteristic of the aesthetic: she chooses them according to what excites and appeals to her, again, reflecting a subjectivist position. She started out in combinatorics in the area of design theory that she describes as being very structured—something she loved from the beginning. Being able to see the structure of an area of research, especially as you enter it, or the appeal of a problem, is to be able to discern the forms, analogies, and relationships, of parts to the whole, for instance. This seems to operate as an aesthetic form of knowing for Sonia.

Sonia's transcript is silent on *her* sense of the generative source of her mathematical activity so we can only theorise about the potential generative characteristic to her aesthetics. Sonia certainly sees her intuition as playing a part in both motivating and guiding her through a problem. She expresses pleasure several times not only at the thought of *getting a result* but also at the experience of seeing the structure and the connections. She seems to appreciate not just a proof, but also the process of getting there: the chain of logic. Sonia also speaks with great enthusiasm about the initial sorting out of the problem, a putting into pictures that gives her a sense of the structure. For Sonia, coming to know is linked with coming to see and appreciate, or even create, structures, connections, and completeness.

Sonia's interview suggests several levels of aesthetic ways of knowing that are intimately tied to intuition and feeling. Her aesthetic responses seem to be accompanied by feelings of excitement and even euphoria. She is happy to let her intuition guide her and willing to leave a problem, knowing that, at some future moment, she will be *struck* by a new thought.

Case Study 5 - Graham

Graham's primary interest in schooling was physics; he switched to studying mathematics in university because he had the feeling that *mathematics was at a deeper level.* Though initially interested in becoming an applied mathematician, he soon got *hooked on* pure mathematics. Graham learned most of his mathematics on his own but was highly influenced by a supervisor who initiated him into the *school of doing topology*. He tends to work on his own now.

Graham has drifted a bit in mathematics, taking an interest in number theory and algebra. With a new confidence in his ability to *wander about in related subjects*, Graham has returned to physics, albeit through a topological approach. And in choosing his problems, he likes to *drift into something*. He finds it very *attractive* when he can participate in conferences in which topologists, physicists, and algebraists come together to explore the *interface* between the various fields.

Graham claims he has a *semi-visual* style of thinking, a statement he later amends to *largely visual and dynamic... even algebraic things*. This style of thinking meshes well with his predilection for thinking in bed or on the bus, which he regards as more serious than *on bits of paper*. He is comfortable working with informal ideas, without all the details, explaining that *it works in a very qualitative way even though you are doing something which is very exact*. He thinks this is a rather mysterious process, the way in which you can get a *hunch or seeing through something without really knowing about the details*. Graham also sees himself as having changed over the past ten years of mathematical work. Whereas before he might have been able to generate better ideas, he now knows the things that he can do, he can *see things much faster* based on experience.

Part of what attracts Graham in his work is the interplay between physics and mathematics, and in particular, the way in which new insights in physics can provide insights into the types of mathematics that could be developed. This might be considered as generatively aesthetic. However, he does not necessarily think or need his mathematics to contribute to physics. He compares the fashions in physics to those in mathematics but argues that the latter are less faddish since mathematicians will develop their topic systematically until they have worked it out.

Over the past ten years, Graham has also modified the way he decides whether to make a result public. He is now much more interested in contributing something he views as *substantial*, one that would interest other people and in particular his community of topologists. This depends on culture: *For me it is important to be embedded in a particular reference group so that I can judge the significance much better.*

Graham finds beauty *harder to pin down* than significance. He claims it is *a motivating force*, and that is does indeed affect his work, but that he is much less conscious of it (than he is of significance): *There is an aesthetic driving force in mathematics but I wouldn't say that I am consciously thinking of it's beauty because I am not.* Although he thinks this is motivational for most mathematicians, he cautions that it is not the only reason why they do mathematics.

He concedes that often, there is *relatively grungy* work to be done in mathematics *which isn't terribly appealing* but which is a part of the whole process. He distinguishes between a *slick* and *non-slick* proof and claims that brevity is preferable to length, but for *purely utilitarian reasons.* Thus, he is rejecting the evaluative aesthetic and points out that there may be more beauty in the details of a long proof then in *a short terse proof that doesn't tell you very much.* Indeed, he does not like the fact that, in some parts of mathematics, powerful machinery is developed that makes it easy to solve a problem but hard to *understand what is going on.*

He does not think that judgments about elegance or beauty are the same for everyone, noting that *any aesthetic thing is very subjective [...] I would avoid saying that everybody recognises the same thing as beauty.* Yet he does insist that mathematicians are interested in *form and structure*, the generative aesthetic, that this is the basis for their work and as such, they *are much closer to the coalface of those kinds of things than other people.*

Graham's interview has much more consonance with Sonia's than with Maria's. He too acknowledges the motivational characteristic of aesthetics not only in his description of the way he likes to drift into problems but also the reasons he gives as to why mathematicians engage in mathematics, and the way fashions develop. When he discusses form and structure as the basis for mathematical work, he is revealing the similarity between the concerns of the mathematician and those of the artist, from a generative perspective. It seems to be this concern that Graham sees as beautiful, or aesthetic in mathematics rather than evaluating the elegance, or not, of proofs. He is certainly aware of a certain discourse around elegance that is present in the mathematical community and admits that he finds some proofs more appealing than others, sometimes because of their brevity but mostly because of their perspicuity—an aesthetic criterion shared by many mathematicians including Hardy (1940).

Graham also addresses another aspect of the evaluative characteristic of aesthetics in his explanation of when he publishes his ideas. Rather than just publishing any result, Graham now likes to make sure his result is both substantial and interesting to his community.

<u>Summing up</u>: There is a much more sophisticated and socialised discourse around the evaluative characteristic of aesthetics, whether or not mathematicians perceive themselves as being involved in decisions about the interest and significance of results. Maria's lack of aesthetic motivation highlights the very different, more aesthetically-driven way in which both

Sonia and Graham select their problems and fields of interest. Maria's very literal way of engaging with mathematics also contrasts with the visual, qualitative, and aesthetically-sensitive way in which Sonia and Graham prefer to engage. In general, it seems difficult to find a language with which to discuss the generative characteristic of aesthetics, though we may catch glimpses of it through the intuitive and affective responses of mathematicians.

NURTURING OF INTUITION AND INSIGHT

Intuition is a complex construct that was investigated in Chapter 5. It seems to operate in many different types of human endeavour, including the artistic and scientific, and yet is still poorly understood by both cognitive scientists and philosophers. Many mathematicians have talked about having sudden intuitions that provided them with a breakthrough in their problems or understanding. However, the very nature of such breakthroughs, insights, or clear understandings, makes them difficult to analyse. As was demonstrated in Chapter 5, many mathematicians draw close links between intuition and aesthetics. In the following, we will not only provide a glimpse into this complex construct but attempt to tease out the different roles intuitive processes play in mathematicians' work.

We choose to offer case studies of three mathematicians, Lance, Sarah and Kyle, who provide a sense of the variety of ways in which the mathematicians who were interviewed perceived and used intuition. Some mathematicians, although not many, simply dismissed intuition. As explained in Chapter 5, a few who were uncomfortable with intuition, were happier using the word insight. However, the majority of mathematicians expressed one of two beliefs about intuition, either that:

(a) intuition is mostly just a function of existing knowledge and prior experience, or

(b) intuition is a way of knowing that is more than just existing knowledge and prior experience. It is an independent epistemological factor.

This second category of belief leads to very divergent ways of describing intuition and its role. However, the basic, binding idea is that intuition is something, over and above prior knowledge and/or past experience, that helps mathematicians know, see, or perceive in a possibly different way than prior knowledge and/or past experience might allow.

Case Study 6 - Lance

Lance is a statistician, *not a mathematician*, who has always been good at maths and who had both parental and non-parental influence in pursuing a career in statistics. He works on many different projects at a time, including traffic modelling and credit scoring, which all require both theoretical research and computer testing. He often participates in collaborative projects and enjoys the complementary skills he and his colleagues offer each other.

Lance thinks in a combination of symbols and pictures, relying highly on the visual when he is working with graph modelling problems. He believes that he knows something when he has succeeded in proving a theorem. His sense of knowledge increases when his theorem succeeds in making it through the referee process. When he knows something is right, Lance claims that it *usually has various properties and symmetry is one of them.* In fact, he believes that symmetry is one of the properties of a good theorem. Conversely, he states that he has reasons for knowing that a theory is right when it is *elegant and beautiful, succinct and well-expressed.* This illustrates the way in which aesthetics operates as a way of knowing for Lance.

Lance begins with a strong view on intuition, asserting that *you need a lot of intuition in order to do research in statistics.* He initially connects intuition with *good understanding,* and implies that his work would be impossible without intuition. Yet, for Lance, intuition is not something that all mathematicians have: *some people have it and some people don't.* After this initial confident start, he begins to re-think his notion of intuition. He changes his mind over the course of the interview from *I might say that I understand something by intuition. But intuition is not a word I would use. Insight, sure, but not intuition because it doesn't describe what I do.* He introduces the notion of insight to differentiate what the mathematics can do from what the mathematician does: *If I am talking about the person I would use the word intuition, if I am talking about maths. I might say that gives an insight into something.* He is presumably trying to differentiate between two types of understanding. The intuitive type is the mathematician's informal understanding of something whereas the insight type, as the word suggests, is a sudden understanding provoked by the mathematics. He might be referring to this latter type of understanding when he describes the *gut feeling* he had with respect to discovering a certain theorem. He describes knowing it was right because he could see a parallel set of results emerging from the new notation system he developed: *there was a gut feeling that we had cracked the language in which we were expressing our concepts.*

Lance's notions of *gut feeling* and *insight* correspond with Hadamard's (1945) notion of intuition. Hadamard (1945: 33) asserted that intuition operates in the "incubation" stage of mathematical discovery, the stage during which no work of the mind is consciously perceived, when things just seem to "click into place". Thus the mathematician will suddenly know something, which will then have to be verified and proved, apparently out of the blue. Lance also alludes to this incubation stage of mathematical work when he mentions needing to have a piece of paper by his bed in order to capture the *good thought* he sometimes has when he wakes up.

Unhappy with his description of intuition, Lance subsequently decides to introduce the notion of pattern-matching better to explain the intuitive thinking process:

> *If you can see the pattern here and translate it to there you might well have new results, new theorems, etc. I would have called that pattern matching intuition, understanding, whatever, because it is a way of predicting the unknown.*

Lance now seems to be attempting to describe his way of gaining new knowledge, and proving new theorems. Pattern-matching is a strategy he uses in problem-solving. It depends on the informal, or intuitive, understanding he already has of certain mathematical concepts. For him, this strategy operates outside the formal knowledge and deductive reasoning of mathematics. It is necessary for crossing those gaps that exist between the deductive steps of mathematics.

It is difficult to tell where Lance stands in terms of thinking about intuition as a function of prior knowledge and experience or as something more. In more or less equating intuition with informal understanding, he casts himself in the former position. However, one senses in his discussion of *gut feeling* that Lance does at times perceive relationships in a different way than prior knowledge and/or past experience might allow. The gut feeling he describes might well have an aesthetic origin, the desire for duality in the parallel set of results, which may in fact provoke an intuitive response.

Lance is strongly aware that there is something beyond the propositional, formal, and deductive that is operating in his coming to know, even if he has a difficult time describing the nature of this knowing and thinking. He relies heavily on an intuitive mode of thinking in his work, playing in the world of the informal and the general, rather than relying on hard properties, theorems, and definitions.

Case Study 7 - Sarah

Sarah is a geometer. She was first aware of her facility and enjoyment of numbers at the age of about four and a half. She describes her highly aesthetic view of mathematics, as how, starting in high school the *pure maths we did just took hold of me... It sounds silly but it was the way things fell into place, structures, connections, patterns, the beauty of it all.* She completed her thesis in differential geometry after spending many years teaching in school and raising a family. She does not see herself as one of those *high-flying mathematicians* nor does she feel competent enough to supervise students. Yet she has an active research career, collaborating from time to time with colleagues and participating in seminars. She is a predominantly visual thinker but will sometimes think *in terms of equations*. She claims that she can move from *imaging in picture, through conceptual analysis into formulaic manipulations.*

Sarah knows something when she can see *that it does fit*. She believes that intuition is the *bringer of ideas*. She relates her own intuition to her visual thinking, describing an example of when she came up with an intuitive idea because *I just pictured it*. In addition to asserting that intuition is needed in mathematical work to solve problems, she insists that it is also needed to motivate and delight: *It is the basis of being a mathematician. You need it to enjoy mathematics, to tackle mathematics*. For Sarah, intuition seems to be related to pleasure and motivation because intuition brings her the broad, connected and aesthetic understanding she finds so wonderful in mathematics: *the way things fell into place.* She describes *knowing mathematics* as seeing that something fits in your mind, suggesting again her desire for and enjoyment in connectedness, and thus her regard for intuition.

Although she does not refer to this phenomenon as intuition, Sarah also describes ways in which she believes her brain carries on work in her subconscious, resulting in *"Aha!"* moments: *You mull it over in the back of your brain without your being conscious of it and somehow the brain decides that it has got there and pops it into the front.* She compares this to the way our brains work while solving crossword puzzles.

Sarah also believes that experience and previous knowledge are part of *thinking mathematically*, but suggests that there are other intuitive ways of thinking. For her, these include using patterns, analogy and structures to understand how mathematics works, echoing the view of Lance. These are all modes of thinking that operate at a different epistemological level than previous knowledge and experience. Though she believes that she *takes* [intuition] *for granted*, she tries to help nurture mathematical intuition in her students by helping them realise *how they are making connections so that maybe they can use this realisation to do more mathematics*. This suggests that mathematical intuition can be learned, provided one thinks about, or attends to, certain types of relationships, a view that is reinforced by Fischbein (1987). (And see Burton, 1999c.)

Sarah does not believe that intuition is reserved for great mathematicians, but, rather, that it is a way of thinking that can be appreciated and developed, and one that brings her pleasure in her mathematical work.

Case Study 8 - Kyle

Kyle is a discrete mathematician who is a *dabbler*, working on and off in various fields including group theory and combinatorics. He was very much influenced by his Oxbridge supervisors and enjoys the social aspects of mathematical work including seminars with his students and collaborations.

Kyle needs to work out the proof to be sure he knows something. It is the *ultimate test*. Later on he admits that this is his *public* view but that *there is a stage before that when I would say it is going to go but it is merely a case of writing down the details.* This resembles Lance's intuitive way of knowing something is true. Yet Kyle insists that this knowledge is simply *based on experience*. Like Sarah, Kyle also enjoys that feeling of knowing how something will work without writing down the proof: *It is a very good feeling, I wouldn't do maths if it weren't for that.*

Twice, Kyle refers to the role of *luck* in solving problems: when he comes back to a problem after having taken a break from it, the ideas will flow and *with a bit of luck something will come out*; and, in describing how he knows which way to go with a problem, he claims that *you have to be a bit lucky*. Both Lance and Sarah called this type of luck either intuition or insight but Kyle prefers to stay away from mysterious processes: *there is nothing mystical about it... it mustn't carry any suggestion that there is some other-worldly process going on.* In fact, he does not even like to use the word intuition, claiming instead that insight is the better word and cautioning that *there are things that go on in the brain that can't be clearly explained but they are processes which have been polished by past experience.*

In the same way that Kyle finds nothing mystical about his insights, he finds nothing mystical about mathematics. He takes a *Chomskian* view of mathematics in that he believes that certain questions are natural ones in mathematics and that *there is some deep structure in mathematics which causes us to do the mathematics we do.* In discussing what makes a research paper publishable, the one area where most mathematicians invoke some kind of aesthetic criterion, Kyle begins by emphasising correctness. In fact, he claims that the thing that distinguishes mathematics from any other subjects is that one must only decide *whether it is right, that is one decision that you make that is not a value judgment.* He thinks that the referee's role is to make sure that *every step follows logically.* Later on, he concedes that aesthetic judgements are involved in deciding whether a theorem is *nice.* He has two criteria: *it mustn't be too complicated to state; and, it should open your eyes to something that you haven't noticed before.*

Even though Kyle attributes his mathematical feelings to previous knowledge and experience, he acknowledges that when you become immersed in a problem, things begin to happen in your mind at a subconscious level: *Things re-arrange themselves in your brain.* Then you see the way that a problem should go *without necessarily knowing how to do it.* This ability increases with experience according to Kyle: *It happens more now that I have had quite a bit of experience with mathematics.*

Kyle's interview exemplifies the disagreement amongst mathematicians over the nature and role of intuition. Kyle is very consistent in his thinking about mathematics; there is nothing mysterious, for him, about what mathematicians do nor how they do it. He does acknowledge different stages or levels of knowing, admitting an informal one which precedes the formal knowledge provided by proof. However, Kyle's view of his informal knowledge, based on experience and past knowledge, differs from Sarah's view.

<u>Summing up</u>: The mathematicians described above all invoke some kind of intuitive thinking in their coming to know, although they differ in their beliefs on the nature of that thinking; these beliefs seem to connect with their more general beliefs about mathematics. Nevertheless, they see intuition as depending in some way on informal knowledge and of being most crucial when they select problems, begin working on them, and when they are judging their worthiness. They tend to use the words intuition, insight, and gut feeling rather interchangeably, sometimes even linking intuition to aesthetic insight and often citing an affective component.

RECOGNITION AND CELEBRATION OF DIFFERENT APPROACHES IN STYLES OF THINKING

We have chosen four mathematicians to represent different combinations of thinking styles that are possible, given the three styles presented in Chapter 4. Our choices were influenced by wanting the transcripts to provide enough detail. We hope that, in hearing their voices together with the detail in Chapter 4, the process by which categorisation was achieved will be clear. To repeat the styles, the mathematicians might use only one of either:

Style A: visual (or thinking in pictures, often dynamic),
Style B: analytic (or thinking symbolically, formalistically),
Style C: conceptual (thinking in ideas, classifying).

Or, they might use two of the three above, providing three different possible combinations (A/B, A/C, B/C). The thinking style they preferred to use depended upon the problem on which they were working. Three mathematicians used all three of the above (A/B/C), again depending upon the problem. That is any one mathematician could exemplify one out of seven different possible positions with respect to thinking styles.

Our four case studies are Barbara, a Style C thinker, Stephen, who uses Styles B and C, Miriam, a Style A thinker and Peter who thinks using Styles A and B.

Case Study 9 – Barbara, a Style C thinker

Barbara comes from a family with members who had been to university but it *wasn't the expected thing*. She was overawed by Oxbridge where she did her first degree followed it with a teaching qualification and then taught for five years. During this time, her husband finished his Ph.D. and they moved to North America for him to take up a post-doc. With no work permit, she did voluntary work with pre-school deaf children, a mathematics programme for ethnic minorities and another for gifted children. As a result of these experiences, she decided to do further study and, having been placed in the top percentile for maths on the preliminary exams, was advised to pursue it. She and her husband returned to the UK where she completed an M.Sc. in statistics and then a Ph.D.. She calls herself a *mathematical or theoretical statistician*. This kind of checkered history, heavily influenced by coincidence, was not at all unusual amongst the women mathematicians who were interviewed (see Burton, 1999a).

Barbara tends to talk metaphorically, remembering at one point in the interview, that *W.W. Sawyer once described maths as like being on a drift of logs and one drifts away and you have to bring it back. I can see aspects of it that are like that.* At another point, speaking of understanding, she says:

> *One's own understanding might actually be wrong. So one's previous understanding might not have been right for where you are now. Picking up a piece and seeing that it should be turned around and put in a different way. And you know that you know something when it is in, but that may be wrong or a new transformation might be around the corner. I have another image of taking the pieces out to look at them, like diamonds, and then putting them back.*

Her thinking, she claims, is not sequential. *It comes all at once, and you can't do it in steps before you know what the steps are.* She does not see *lines of writing with the proof written out*. Neither is it pictures but *there is some pictorial element but I don't know what it is*. This description is typical of those who were categorised as Style C. They often spoke about identifying the important concepts in their problem and placing them into bubbles in a bubble diagram *a picture, but not a mathematical picture*, which could then be manipulated and the results observed. Barbara expresses a very personal relationship with the objects of her thinking, seeing *numbers in colour*. She says:

> *When I am inside my head, I am thinking about these things. I know what they are but I am not seeing them. I am trying to get inside the formalism and see what it's limits are...It's stepping back from the formalism and*

trying to put some distance to see if there is another way around it, or can use a different idea.

Case Study 10 – Stephen, who uses Styles B and C

Stephen's route into his current work was very complicated. He began his career as an apprentice in an electronics company, moved on to be a trainee accountant and then did some computer programming. At that point, he decided to get formal qualifications, having to gain school certificates in order to secure entry to a university. After his first degree, he worked on a funded research project in a new area of electronics and then went on to get a Ph.D. after which he entered the field of optimisation of mathematical simulations and subsequently headed a project on supercomputers. It was after this that he got his present position in a university teaching operational research and statistics. This circuitous entry into mathematics was not unusual for the females interviewed, as in Barbara's case, but was most unusual for the males.

Unlike Barbara, Stephen does work sequentially, *trying to produce a taxonomy. In mathematics, you solve one particular problem which leads on to another, etc. and then you look for generalities.* Stephen is an example of a number of mathematicians who seem to feel that thinking in pictures is the way they *ought* to be thinking, but it does not come naturally to them. He says:

> *I try to think in pictures but often I don't find them very helpful. I am not a very visual person. I think in terms of equations but I think it is problem driven. I don't think I can ever deal with solid geometry problems. If I am looking for a taxonomy, I use hierarchies, classificatory pictures.*

Stephen is categorised as utilising Styles B and C.

Case Study 11 – Miriam, a Style A thinker

Miriam was good at mathematics at school, and achieved. Her parents did not go to university but were both *good at maths and enjoyed it at school* and they encouraged her, as did a teacher who was important to her development and career choice. Another Oxbridge graduate, Miriam became a particle theorist but moved into quantum gravity which, she felt, combined particle theory and relativity.

Miriam describes her thinking style as follows:

I see this picture of various tori, with meridian disks. I am thinking of a problem that comes up from one of my research students. I have been trying to get to grips with some abstract concepts. I find it hard to take in very abstract things...I have these visions of these diagrams and see how things fit together and how you can identify, handle, bodies to make particular manifolds. The diagrams are static. I am thinking about how you can put them together, identify surfaces of different ones to fit them together, how you can label or 'colour' (by assigning numbers) elements from a set...By labelling the different regions you can write down an expression which should characterise the topology that you have got.

Miriam appears to be a classic Style A thinker and her language is permeated by the use of the verb 'to see':

I start to see the way things fit together; It is very exciting when you make a connection, put a piece in a jigsaw, see how some ideas that were understood by one group of people were exactly what another group of people needed to make something work; I suppose I am interested in seeing how things fit together, seeing patterns in things.

Case Study 12 – Peter who uses Styles A and B

Peter's father was a mathematics teacher, *there were lots of maths books around the house* and he was encouraged by a teacher in his *random comprehensive* to undertake a mathematics degree. He also went to Oxbridge which was *a different world* and stayed on to do a Ph.D. He now works in analysis, doing *two quite different sorts of things...Inside functional analysis I do cohomology theory and harmonic analysis. The complex analysis is different.* Peter is one of the very few mathematicians who prefer to work on their own. He says:

I think mathematicians as a whole don't seem to work together. I do prefer to work alone. I don't particularly like working with others. But sometimes I do work with others, but I don't seek it out. When it happens, it happens by chance. But collaborations are a combination of individual work in that we have different knowledges, different backgrounds, and we bring different things to the work...I don't understand what people who have almost identical knowledge get from working together.

Peter's view is strongly disputed by many of the other mathematicians interviewed (see Chapter 7 and Burton, 1999b).

Peter's description of his thinking is graphic and leads to his thinking style being categorised as Styles A and B.

Typically I do it at night in bed. I have an infinite board in front of my eyes and I draw pictures on it. It is about 2 feet before my head and I write on it with my finger. I read it, do calculations...The pictures I draw are generally static. They are usually diagrams. When they are not equations, they are diagrams of graphs in the maths sense, vertices and lines, three-dimensional and four-dimensional versions of these things. I draw them, draw on them, add to them, move them, often things get stretched out and turned like a Cat's Cradle. I might imagine myself taking two parallel lines with a rubber band in between and I might stretch them and move them and see what happens. But generally it is the nature of the maths that nothing usually moves, rather than the way I think about it. I think, principally, I imagine algebra and write down the equations. I do some sort of calculation. I could quite easily go on for five hours doing this in my head.

Summing up: As indicated, thinking styles are both highly personal and very influential in the ways in which a mathematician comes to know their mathematics. Mathematicians seem to manipulate their working environments and their problem approaches in order to match these with their preferred thinking styles. (For similar results in engineering, see Cowan, 1983) Their apparent attachment to these thinking styles may be connected with a (pleasurable) sense of identification that they feel between the mathematics and how they think about it. The fact that such thinking styles are not acknowledged in classrooms is a concern that was expressed in Chapter 5 and is addressed in Chapter 10.

THE GLOBALITY OF ITS APPLICATIONS - CONNECTIVITIES

This category of the model was the one on which all the mathematicians agreed. For some, connectivities were outwardly-oriented, that is, connections to the real-world, to real problems, to physics or other disciplines. While for others, connectivities were inwardly-oriented, that is, represented connections between the different fields and methods of mathematics. The following two case studies of Carl and Donald illustrate these two perspectives, and also show how they need not be mutually exclusive. They also attempt to explain the epistemological dimension of connectivities.

Case Study 13: Carl – outwardly oriented

Carl is an applied mathematician who works primarily in mathematical physics. He has always been intent on pursuing a research career and has been interested in science and theoretical physics since childhood. He works very closely with colleagues, partly as way to stimulate his ideas, to *think more fruitfully*, and partly because he feels uncomfortable with the *tremendously savage competition* in his field.

Carl thinks in an *ill-defined pictorial way* and moves back and forth between formal manipulations and visualisations, which are often more dynamic than static. He works with an *abstract picture* of what is going on and then *fills in the details later on.* He claims that his way of working is a *function of experience.*

In the past Carl's area of interest was *fairly narrow*, making him feel that he wasn't doing *anything particularly useful or significant.* Currently, he is attempting to change that. Although he sees his work as hovering *on the edge of applicability,* he would like it to have more applications. He realises that some areas can have physical and technological applications but that they can also be interesting from a mathematical point of view. Since he thinks of things from a mathematical, a more *fundamental* point of view, his work is not immediately geared to applications, but Carl would like to *establish the bridgehead which allowed that.* He adds that he *is always keen to try that*. He laments the fact that most people do not see this as a goal even though there is a *rhetorical attempt* to make applications and that is it is not easy to make this bridge that connects all the way *from theoretical ideas to applications*. Carl also laments the increased specialisation in mathematics that can prevent people who are interested in a certain problem learning about what others, in different fields, may have to contribute to that problem. This may result in a reduction of possibilities for connectivities, that is for connections and applications between different areas of mathematics.

Carl sums up his desire for connectivities in describing what knowing means to him:

> *you see that various things are connected in a way that you hadn't seen before. Making the connections is what you are trying to do. You are trying to show that your techniques can solve a range of problems that are connected and you may not have realised before that they were connected.*

Case Study 14: Donald – inwardly oriented

One might think that applied mathematicians are more strongly disposed to connectivities than pure mathematicians. But Donald, a geometer who thought of himself *as a pure mathematician from a very early age* also speaks of the importance of connectivities. Donald first appreciated mathematics as a *way of doing creative and technical things without actually having to tackle the practicalities.* He continues to find pleasure in this aspect of mathematics, that you can *do all these interesting and exciting things without having to go out and do things with them.* He thinks of pure mathematics as a field where *things are done very carefully* and applied mathematics as a field *where they don't care about things like that.*

Donald is a differential geometer but he has broad interests. He is against *splitting off into specialties* because he thinks that *it produces a lot of marginal work.* He believes that *all specialist work becomes sterile after a while* and that *staying in the mainstream* is a good way of making sure that *you might do something interesting.* He sees mathematics as being made up of *grand traditions* that provide a *broad front* within which worthy things can be done. Only someone with *exceptional talent* could achieve anything worthy by going off on *some specialised little track.* Donald is expressing a strong disposition towards connectivities; working within the grand traditions assures connections with the body of mathematical knowledge—and this connectivity can influence judgements of worth or significance.

Even though there are parts of mathematics he knows nothing about, parts where he *cannot really see any connection*, what he really likes is *to see connections for example between differential equations and geometry.* He believes that mathematics is *like a map with roads between different areas* but that every mathematician has a different map that extends into regions that *seem murky* to him, regions he doesn't *fully understand.* Therefore, part of understanding for Donald, is seeing how something connects to his road map of mathematics, how it connects to geometry or to number theory.

Donald is also interested in doing work that has some application. He says that would make him *perfectly happy.* Where he may have once shunned *applicable mathematics*, he is now interested in working on areas which *have some contact with reality.*

Summing up: Carl and Donald, although they came from different parts of mathematics, shared a desire to make connections. This was the case with almost all of the mathematicians interviewed. Connectivities proved to be a powerful motivator for mathematicians to work, even if their work was not located in overtly connected, or likely to be connected, areas.

THE WHOLE MATHEMATICIAN (EPISTEMOLOGICALLY SPEAKING)

We end with three mathematicians, two females and a male, and we tell their stories holistically, that is from the perspective of all five categories that constitute the epistemological model of mathematics, its person- and cultural/social-relatedness, aesthetics, intuition, styles of thinking and connectivities. We are doing this to introduce the person-ness of the mathematician to this book's story, but also in order to demonstrate the extent to which the general model can effectively reveal and describe the specific, local nature of mathematicians' claims about how they come to know mathematics. (References in this section to the categories of the model are in bold).

Case study 15 – Elizabeth

According to her colleagues, Elizabeth is a theoretical physicist of some considerable reputation. She claims always to be *fascinated by science and interested in physics* and remembers reading, in the Sixth Form (at the age of about 16 or 17) a book that focused her attention on particle physics. During the summer vacation between the first and second year of the Sixth Form, she read as much as she could about particle physics and this set her on her career trajectory. In her final undergraduate year, she was offered a project that she found *fascinating* and while doing this work, she found that she *was much more interested in the theoretical background and had to learn a lot of quantum field theory to understand it*. In order to complete the project she was *essentially teaching myself* and her *project contained all the results and all the theory as well*. The required self-reliance *probably started at school* where there were teachers who *really believed in the comprehensive ideal and they were probably quite well qualified.* There was no academic background in her family and, although encouraged by her parents, *they couldn't help me*.

It is the natural world that excites Elizabeth about physics. Her supervisor for her undergraduate project said that it was *obvious that I would get a first. It hadn't been obvious to me. He also commented on being able to do research in the summer time.* This made Elizabeth realise that, at that university, research was marginalised. She therefore chose a university for her Ph.D. *that was higgledy-piggledy but was full of research*. In speaking about her training as a theoretical physicist/applied mathematician, Elizabeth uses words such as *exciting, stimulating, collaborative* although she also speaks of much less productive experiences. She notes the importance of *not competing* and also of *discussing with people* [*who*] *are open to discussing with me*. She has professional experience in a range of prestigious institutions in the UK and abroad, which points to the value of her work in her area. But despite this, and the encouragement of her Head of Department, she thinks that *maybe a personal chair would be over-reaching myself.*

Elizabeth can be seen to have a discursive and collaborative attitude to her research, making it clear that the **personal *and* the social** are important to her. She reveals the importance of the social dimension when she describes how she likes *discussing ideas because I find it very stimulating and I also want to share my ideas.* The personal dimension emerges more subtly when she explains how she begins thinking about a problem:

> *If I am interested in a problem, I start to wonder what is going on and, if I think about it and write down the underlying theory and work out what is going on, alright so that is understood. It means I can now see how something works. This is the underlying theory...The empirical data is not fundamental to my understanding. Particle physics should explain the nature at high energy or in the very early universe and one does have to confirm that with data but I am not interested in that data fitting, nor in the very abstract mathematical side but in inventing and developing things to see how to extract the fundamental understanding. I am a particle physicist but I am working at the overlap between particle theory and cosmology. Sometimes it is easier to talk about the early universe than it is to talk about particle theory.*

This demonstrates how personal, particular and focused, are Elizabeth's research interests, as is the case with most of the mathematicians interviewed. Far from having wide-ranging interests and very general mathematical strengths, they freely speak of not understanding or not being interested in certain aspects of their chosen field, never mind of mathematics more broadly. But the social exploration and interchange are central to Elizabeth's expressed research needs.

The quote above exemplifies how Elizabeth's choice of problems has an **aesthetic** dimension. Yet aesthetic criteria also feature prominently in the expectations Elizabeth holds for publishable results: *elegance is fundamental.* Similarly, aesthetic criteria are also central to her judgements of the fitness of her results:

> *Anything that I am working on I would want to look elegant enough that it could actually describe nature...[So,] I would be looking for something about generalisability and not a potential with too many parameters that nature would never describe that. You might be able to write it down mathematically, it might obey all the axioms of quantum field theory, but it had the parameters in that the world is not like that – clumsy! ...If*

> *something were inelegant or not beautiful, too contrived or wrong, I would dismiss it. So elegance is fundamental to getting going.*

Aesthetic appeal for Elizabeth, as with many of the other mathematicians, is not only about simplicity but also about economy.

In discussing how she thinks about a problem, Elizabeth says that she is:

> *using an intuition I have developed of how the fundamental physics works out. I am trying to understand the problem but to think if the theory is symmetric, these fields have to be like that because those fields are like that. Trying to understand what we should be looking for.*

For Elizabeth, therefore, the **intuition** is the starting point of the process. Things can become *obvious* but *not so obvious that it is easy to work out.* She explains that, as is generally the case (see Burton, 1999c), *my intuition is a result of a synthesis of previous results that I have shown or putting it together with what other people have shown and applying it to this particular case. I am synthesizing my previous experience and other's.* That is, **intuition** is a combination of knowledge and experience whether one's own or other people's.

Elizabeth describes her **thinking style** in terms of how she represents the problem on which she is working using words, and claims that:

> *I don't seem able to visualise. I don't think in terms of equations. I might know the solutions to equations or the solutions in special forms. I would know about behaviour. I might think in terms of Feynmann diagrams, which are pictorial representations of the interactions in a quantum field theory. Some people seem to be able to think in terms of colours, or visualise geometry. I can't do that. What I can do is visualise what the elementary particles should do in terms of results, decay, interactions.*

Elizabeth is differentiating between a more conventional use of visualisation in terms of pictorial representation, and her use in which what she *sees* is the behaviour of the objects in which she is interested. However, like many of the mathematicians, Elizabeth claims that her *thought processes are sometimes a bit of a mystery to me because when I am doing anything I spend a lot of time just thinking* but not reflecting on the thinking process. Her explanations for how she goes about that process, cause us to place her in category C as a classifier (see Chapter 5).

When talking about publication of results, Elizabeth wants to see *novelty, something that hasn't appeared before theoretically, perhaps a new mechanism for understanding something* but she also thinks it is important that areas should *overlap. I am very interested in connecting up areas. Applicability across fields is important*. So, **connectivities** are also a feature in the expectations Elizabeth holds for publishable results.

It can be seen, therefore, that the five aspects of the model can all be read in Elizabeth's discourse, not because, in the interview, she was led in those directions (see method discussion in Chapter 2) but because they constitute the way in which she conceives of and articulates her epistemology.

Case Study 16 – Alexander

Alexander also *always wanted to become a physicist rather than a mathematician* because of an aunt who was an important influence. But, because he did well, he took an undergraduate degree in mathematics while still keeping open the possibility of a research career in physics. As a result of a number of coincidental external experiences, but also because he found pure mathematics easier and was stimulated by a particular mathematician who became his Ph.D. supervisor, he chose topology. He has been an important contributor to his field but modestly claims *I do what I can*. His interests extend from topology to computational group theory, geometry and computing. In recent years, he has also developed an interest in making his teaching more accessible to students.

Alexander speaks of his **personal, social and cultural** perspectives on mathematics. He refers to the *different culture* that exists between distinct groups within a disciplinary area. *There is very little overlap between computational group theorists and topologists. If I am writing for the one group, I do not write in the same way as if I am writing for the other group.* He is also clear about the role of hierarchical power, pointing out that *top mathematicians* often accept *dramatically new things and then the smaller communities have to adapt themselves and, for example, learn the new language*. He explains that:

> *you work in a certain sub-community of the general mathematical community. That small community has a common feeling about what kinds of statements you can stop at and what needs to be formalised. And these might be different from one sub-community to another. A statement written out at great length might be thought reasonable in one community but not in another.*

Learning these game rules is part of the socialisation process through which young researchers go as they experience what is un/acceptable within their community of practice. It comes as somewhat of a surprise to those outside mathematics to find that the discipline is both highly fragmented and that each sub-community is as territorial as described. The spirit of competition mentioned elsewhere permeates the research culture, and the language used. For example as explained in Chapter 8, work was spoken of as *trivial*, results were *obvious* or a paper was *significant* and not simply *interesting*. But Alexander explains that, although:

> *the ultimate test of whether something has been proved or not is commonly taken to be that the proof you have given is capable of being formalised, to formalise a proof without diagrams is much easier than to formalise a proof with diagrams [and] a lot of this formalisation is comparatively low level. Mathematicians don't bother to do it because we know it could be done but we don't do it...you cannot formalise completely and you have to make a judgment about where you stop. Topologists and geometers are used to stopping long before group theorists and people who do things algebraically. So, if you get an algebraist looking at a topologist's proof, they are likely to say I don't really regard this as a proof.*

Contrary to expectations, therefore, there is nothing 'objective' or 'sacrosanct' about a proof. It is what the community to which it is offered accepts, written in the formalisation that is the acknowledged norm. Amongst many mathematicians who have written about this see, for example:

> When the mathematician elaborates rules of logical coherence, rules of exclusion – in short a formalism – he [sic] is constructing a universal language that permits him to recognize properties of the object he constructed in the first place. In the end he 'discovers' only the consequences of what he himself has conjured up!. (Changeux & Connes, 1995: 20)

> The formal-logic picture of proof is a topic for study in logic rather than a truthful picture of real-life mathematics. (Hersh, 1998: 57)

> A *proof* is something informal, or in other words a product of normal thought, written in a human language, for human consumption. (Hofstadter, 1979:195: original emphasis)

> Mathematical knowledge resembles *empirical* knowledge - that is, that the criterion of truth in mathematics just as much as in physics is success of our ideas in practice, and [that] mathematical knowledge is corrigible and not absolute. (Putnam, 1998: 51: original emphasis)

> Insofar as the search for satisfactory Foundation theories aims at any absolute criterion for truth in mathematics or fixation of mathematical method, it appears doomed to failure, since recognition of the cultural basis of mathematics compels the realization of its variable and growing character. (Wilder, 1998: 197/8)

Although Alexander alludes to **aesthetics** in his explanation of the way in which work is judged by the community, he addresses it more specifically when he says that *the simple ideas are the ones people say are beautiful. Simple ideas that are powerful are the ones I think are beautiful. They can be used to prove surprising consequences.* Alexander recognizes that, within the community, there is an understanding of what is aesthetically appealing (*simple ideas*), but he has also developed his own conception of aesthetic appeal, valuing not only what is powerful, but also what is surprising.

Like many of his colleagues, Alexander believes that **intuition** is informed by both mathematical knowledge and experience but that there is more to it than that. He begins his discussion of intuition with a description of an *example of something that is very well known but I didn't know it. I knew it as well known and it was connected with my teaching.* He says, acknowledging that some intuitions may be inadequate:

> *the standard geometrical intuition is really relevant whereas with higher dimensional problems you have to use two or three dimensional intuition but it has to be guarded because things don't work the same way. So the intuition has to be informed by mathematical knowledge and experience...The way you use intuition, you know which aspects of everyday objects are applicable in this more abstract situation.*

Alexander provides an example of how a mathematician thinks concretely about abstractions:

> *If you take two pieces of paper and try to imagine them passing through each other, they intersect along a line. Every mathematician knows that if you do that in four dimensions they meet in a point. In five-space, they won't meet at all. So you can still think of these planes as pieces of paper but with unusual properties. If you have a three-dimensional space meeting a two-dimensional space in four dimensions, they will meet in a line. If you are thinking about a complicated situation, for example with objects intersecting each other, how do you think about this? You think about it as objects intersecting each other. That is intuition. To think of a three-dimensional object as being two-dimensional, how can that be knowledge?*

However, he warns that intuitions can also be false and need to be checked:

> *For many mathematicians, your mind needs to get hold of something concrete and, when it does that, it is going to be telling you something false unless you pay attention and use what you know as mental annotations. When I have two three-dimensional pieces of surface, in four dimensions, I expect it to meet in two dimensions; that is intuition, it is not the result of*

> *reasoning. You cannot prove that they meet in a line. It is not a theorem. Intuition is what you expect to happen and the basis of what you expect to happen is your work but you end up with something much better, even if the work was not distributed equally.*

These themes around intuition, of it resulting from the combination of knowledge with experience but operating over and above them, of the inadequacy, incompleteness or even falsity of intuition, and of the links between the concrete thinking of the mathematician and their intuitions are pursued in many of the interviews (and see Chapter 5).

Alexander's **thinking style** is *a kind of visual picture* but, depending on the kind of problem, it might be static or dynamic. His language is visual. He speaks about what he *sees* and how he tries *imagining yourself as a tiny little insect in this huge {mathematical} structure. If I can do that, I find it helpful. But I don't do it that much. I tend to think of things more my size or smaller, things I can manipulate.* This description fits the classic visualiser, i.e. thinking style category A.

As with many pure mathematicians, for Alexander **connectivities** can be expected even though they are not necessary to his work. He points out that: *Ideas from one area often do have relevance to another area but that isn't one of the more important things for me.* However, he goes on to speak about some current work where he wants *to see how things develop for their own sake.* But, he *would be very pleased if it were applicable to some real life things.* However, there are limits to his desire for applicability:

> *probably the most relevant area would be some kind of military application and I don't know what to do about that. I hope it doesn't actually come about. I do choose collaborators who will refuse to take military funding partly because of this worry. I would feel uncomfortable being on a collaborative grant with people who wouldn't mind.*

Alexander is one of very few of the seventy mathematicians to answer directly the question, *How do you know when you know?* Most of the mathematicians responded by invoking either the jigsaw metaphor – putting in the last piece, or a geographical, or journey, metaphor (see Chapter 7 and Burton, 1999b). Alexander, on the other hand, makes a statement of connection. His process of coming to know is to follow a path backwards and forwards between something that is for him a certainty and that formalisation which is the object of knowing. He says:

> *I know if I think I can formalise it. If challenged, I could give an answer at the next deeper level until I get down to the axioms of set theory. I might be mistaken but that is what I think. There are other statements that I know; for example, I know that every three-manifold can be triangulated but I*

have never read the proof. I have just been told it so many times that I accept it by authority. In my research, I like to be able to feel that I could go back all the way.

As with Elizabeth, Alexander's acknowledgement of the five categories of the model throughout the interview is unprovoked, although he describes them within the particularities of his own mathematical practice. His interview transcript lends additional weight to the view that the model is a justifiable description of the epistemology of mathematicians, but reinforces the view presented elsewhere about the heterogeneity of the positions held by the different mathematicians (Burton, 2001b).

Case study 17 – Kathy

No-one in Kathy's family had ever been to university but she won a scholarship to Oxbridge where she had to learn to be *extremely self-reliant and it was a hard lesson.* After graduating, she did an M.Sc. in Mathematical Statistics and Operational Research and, subsequently a Ph.D. in a similar area. Looking back on her Oxbridge experience, she says: *ultimately my learning experience was alright but I could have drowned.* She describes her area of research as applied probability.

Perhaps because of the area that she is in, her approach to mathematics is one that embraces uncertainty. Knowing means understanding:

how a particular system is described mathematically and what the mathematical implications are. If I solve an equation, I have got an answer. I know it's an answer because I have followed a process and got to an end-point: given, construction, proof.

The view Kathy expresses here is very sequential and logical in contrast with the gestalt of which other mathematicians have spoken or even the recursive filling-in and fitting-in that would fit with her style of thinking. She continues:

Sometimes you never know if you know. You can do it by contradiction, by finding fault. You can never know it is right. You always have to live with uncertainty. You don't ever know anything. So, knowing is feeling fairly certain that I have worked through the process correctly and the end-point I have got is correct, I didn't make a mistake somewhere in my application of the process or my understanding of how the process should be working. I do frequently feel very sure that I have done it right and sometimes I subsequently find I haven't.

So, sometimes even when she has appeared to work correctly through the process, she can get it wrong. What does she do when she finds that she has not got it right? Describing her resolution of such a situation, she says: *it became alright. I changed the boundary conditions* - returning to question the givens is a classic way, in mathematics, of making necessary adjustments that allow the process to continue. A very interesting contradiction is represented in the passage above in which, at one and the same time, Kathy recounts a very formal 'given, construction, proof' way of solving problems together with a recognition of 'working through the process' and, at the same time, she acknowledges just how little reliability can be placed on such methods by mentioning, in **personal, social and cultural** terms, the uncertainty and the need for choices and decisions that might not be strictly logical. This theme continues in the interview.

For example, Kathy believes that mathematics *is a language in itself* and that she thinks in *non-verbal concepts*; this is what inclined us to categorise her as having **thinking style** C as a classifier. She is insistent that it is not the mathematical objects she is talking about but the *ways of doing things. Rather than the next line coming into your head, it is what you have to do to get the next line,* which suggests that she knows where she is going, what her endpoint is, and the challenge is to find the tools and strategies that can help bridge the gap for her. Although she does say *When I am thinking of a problem, I wouldn't commonly think about how to get from one line to the next.*

She continues *very often you are trying to make connections,* again, possibly, a recognition of bridge building. She also believes that **intuition** *is a natural part of mathematics* and that *to my way of working intuition is fundamental.* Here, Kathy appears to be acknowledging her own person-ness, the degree to which she cannot rely on the formal approach or the process to lead her to her required endpoint but has to use more ephemeral ways of working. Indeed, she says: *It kind of comes, it is like you suddenly have a bit of music in your head, it is almost outside of your control...there are these divergent and probably intuitive leaps around or ideas that come out of the air.*

Later in the interview, she launches into a discussion about **aesthetics** that begins with her saying: *If it was a very elegant way of doing things I would be inclined to forgive a lot of faults.* She continues:

> *Mathematicians like to see ways of doing things which are short, a nice structure, a well-crafted and aesthetically pleasing way of expressing. Clearly, there are very subjective aspects to what pleases you but we are also taught what is aesthetic. A group of people try to induct you or persuade new members that certain things are aesthetic.*

Kathy is making clear how, she believes, through the formal learning processes, the community creates, and inducts new members into, their cultural expectations. But she continues:

> *I am not sure that there are things that everybody would agree are beautiful. Language is culture specific so it is unlikely to be true that people will agree across cultures and times. I don't believe that beauty or aesthetic considerations are much different from other areas of art.*

Here, Kathy is articulating a subjectivist/contextualist view of the location of aesthetics, in contrast with many mathematicians (such as Hardy, 1940, for example) who see the beauty as located *in* the mathematics.

Kathy sums up her approach to mathematics by saying: *Some of the time I am a Platonist. As mathematicians, we want to believe that it is objective but I guess rationally I think it is socio-cultural and emotionally I want to believe it is objective.* This is an honest appraisal of the contradiction in which the mathematician is caught, recognising the normative 'objective' stories told about mathematics but, at the same time, very aware of the personal and social struggles that affect her own results. Kathy's interview demonstrates that she has thought about, and developed a consistent argument with respect to all five dimensions of the model. She raises issues about intuition, aesthetics and connectivities and places these in the context of a personal and social-cultural perspective on mathematics.

SUMMARY

These seventeen mathematicians have provided an opportunity to interpret, refine and particularize the categories of the model through their individual stories. Not only do they highlight patterns that are characteristic of mathematicians coming to know, patterns taken up and explored in other chapters, but they also offer opportunities to tease out some of the differences within those patterns.

The distinction that was made earlier between an epistemology of knowledge, and of knowing, is one that is reinforced by the mathematicians introduced in this chapter. Many of them do adopt a belief in the knowledge of mathematics being objective and absolute. At the same time, they hold an epistemology of knowing that conforms to the model. However, this mental flexibility also enables a tight hold to be kept on the absolute nature of 'truth' and the power and legitimacy of the objects of mathematics. In Chapter 9, the influence of the wielding of this power on the lives of female mathematicians is described. The interview data, exemplified through these case studies, validate the conjecture supporting the model that mathematical knowing is a product of people and societies, that it is hetero- not homogeneous, that it is inter-dependent with feelings especially those attached to aesthetics, that it is intuitive and that it inter-connects in networks. The implications for mathematics education of this heterogeneity are explored in Chapter 10.

In the next chapter, we find out how the mathematicians practise their craft. Where do they find the problems on which they work? What are their working practices? Further surprises are in store as we hear what they have to say about collaboration and competition.

7

Mathematicians' practices – Finding their problems, Collaborating/co-operating and Competing

In this chapter, I examine how the mathematicians go about finding and working on their research problems. Having found a problem on which to work, is that work done alone, as the stereotype of mathematicians would have us believe, or with others? What does this say about the cultural climate within mathematics? What role does competition play in this? This chapter answers these questions.

THE MATHEMATICAL PROBLEMS

When we began that part of the interview that concentrated on their research, I asked the mathematicians how many research problems they worked on at the same time and where they 'found' these problems. Their responses ranged from: *There are lots of unsolved problems in my area,* to the precision of this mathematician who said: *At the moment, I am really only thinking about 2 or 3 problems but one of these is a really big problem which I have been working on for twenty years and I have recently made just a bit of progress.* The majority worked on between two and six problems in parallel although one described the process as: *I have twenty or so problems on the go, all in my head and you pull a different one to the front or it pops to the front.* Some spoke of *'several'* problems or, making a value judgment, *'too many'*.

Where do they find these problems? One said: *The number of problems that fly around on e-mail has enormously increased.* But an obvious place to find something interesting, in someone else's work, was described in this way: *The problems are usually twists and turns of other people's problems. You might read something and wonder if you can take it further.* Another mathematician found problems in *something I hear at conferences, through people you know, reading the journals; for me attending presentations at conferences has been the most useful source.* This eclectic approach was also described by another participant: *They pop up everywhere, talking to colleagues, reading literature, you recognise a setting maybe, or a technique which you have been using elsewhere, or some structure which looks vaguely like a design to which I might be able to apply some design theory.* The sad story of academic pressures lay behind the comment:

> *I think what I regard as my favourite project is so filed back on the list of things to do that it has sort of faded. It originally came from a mistake in another paper where the authors had glossed over something and there was a whole different structure there which came to light during an intensive collaborative visit and that was a real joy finding a whole new structure that hadn't been found before.*

The process of picking up and putting down, revisiting, already mentioned in a quote above, was described by many of the mathematicians. One said: *I have many problems in some areas so I might work intensively on a given problem and then reach a point where I leave it aside and work on something else and then revisit it later.* The decision to move to another problem might be provoked by becoming stuck on the present problem. Another said: *If you have enough things to move to, when you are stuck you leave them alone and walk away to something else.* However, the decision as to when to walk is influenced by why they want to walk. He continued: *What I don't walk away from is a contradiction. If you see two things that seem to clash or push in the wrong direction, it is very dangerous to walk away.* (Being stuck as a natural part of working on a problem was discussed in Chapter 4.)

Although it might appear that the search for problems, and their pursuit, was straight-forward, this was not the case for all the mathematicians. One of the younger participants expressed these concerns:

> *I worry a bit that I might get a problem that I don't know how to solve and that would be unsatisfactory for me, and my employer. It is nice being able to work on what you want but there is also a worry that you might run out of good problems to work on. There are still plenty of problems but I wonder if they are getting more obscure. I do worry a bit if the mathematics I do is going to please me but be useless to everybody else. Although I don't feel that every piece of maths I ever do needs to be useful, it is quite nice to have the feeling that it is*

useful. But there are some parts of pure maths that are really well off the beaten track and although you can't predict what will be useful later I would be prepared to put a fair bit of money on some bits not becoming useful later.

The public image of the lone mathematician was certainly one which I shared before I began this study. But the participants made clear to me just how mistaken this image was. So, having found a problem on which they wished to work, what were their working practices? In the next section, I examine what they had to say about these that were, in their terms, predominantly collaborative.

COLLABORATION AND CO-OPERATION

Much to my surprise, the mathematicians explained that it has become the norm for them to work together, in a style that they called collaborative. This is not at all the same thing as bringing some results to a colloquium or a seminar for the critique of your peers - a practice with a long history. Nor is it the same as seeking out a colleague to comment on and/or annotate an argument offered in the course of solving a problem. Again, that practice has a long history. They were not talking about occasional or haphazard contacts. Collaboration, according to the mathematicians, is when a team of at least two, often more, works on a research problem, frequently with other mathematicians but also sometimes with those in cognate disciplinary areas. The final product, the written paper, is a blend of the work of all those contributing.

As a result of listening to their descriptions, I am now discriminating between co-operation and collaboration. It is frequently the case that a mathematician works, usually in parallel, with those in other areas or other disciplines; the others may provide the datasets, for example, and the mathematician processes their data; this I am calling co-operation. The following mathematician gave one example of such a co-operation:

That involves a huge number of different tasks. There are two people here working for me, there are two others in a neighbouring department, four people in The Hague collaborating and a main client for this work is the Dutch Ministry of Transport. All of this is taking place under the European Framework programme. We had a meeting in Venice last week where there

> *were 20 partners represented. The work is at various levels. There is theoretical research, computer testing, all sorts, all on the go at one time.*

However, many of the mathematicians claimed a style of working which was truly interactive and non-differentiated and where, by the end of the process, that is the submission of the written paper, they could not recall who had contributed what. It is this that I am calling collaboration. Of the seventy mathematicians interviewed, only 4, 3 men and 1 woman, claimed neither to co-operate nor collaborate but to do all their work alone.

The stereotype of the lone mathematician, working in his (usually) study for years on end – which, as I indicated in Chapter 2, I incline to think of as the Andrew Wiles[1] effect – is certainly shared outside of the world of research mathematics. Even within it, I did meet one mathematician who said:

> *I think mathematicians as a whole don't seem to work together. I do prefer to work alone. I don't particularly like working with others... I don't understand what people who have almost identical knowledge get from working together.*

However, even this mathematician said:

> *Sometimes I do work with others, but I don't seek it out. When it happens, it happens by chance. But collaborations are a combination of individual work in that we have different knowledges, different backgrounds, and we bring different things to the work. So the work naturally breaks up.*

This participant, in my terms, was an occasional co-operator. Those who co-operated were often statisticians or pure mathematicians whose knowledge and skills were put to good use by their colleagues in other disciplinary areas.

However, this was not true of all such, and the notion of teamwork in which the sum of the contributions was expected to be greater than any individual might accomplish alone, to which I am giving the name collaboration, was widespread across the different parts of mathematics. It was not only true of the style used to address the work itself, but often also of the manner in which the publication, which announced the results to the mathematical world, was derived. A collaborative mathematician explained: *We both work on the same*

[1] Andrew Wiles, the mathematician who devised a proof for Fermat's Last Theorem, was the subject of a television documentary made by Simon Singh for BBC Horizon. In this, Andrew Wiles was shown working on his proof, in an attic, implying on his own, for ten years.

thing, have different thoughts, send messages. By the end, it all will have got muddled up and you won't know who was responsible for what. How and why did this surprising shift come about?

A CULTURAL SHIFT

> *There was quite a bad atmosphere in research mathematics across the world, of the best mathematicians trying to show they were best. One-upmanship and stuff like that...I think there is more a feeling that co-operation is important now.*

Even though the societal images around mathematicians and their work seem to have remained the same for a very long time, it would appear that the cultural practices within mathematics have shifted considerably. Indeed, one mathematician, who was short-listed for a job interview shortly after we spoke, was concerned that the interview panel would be critical of his curriculum vitae for containing too *few* joint publications. The mathematicians did not offer reasons for this shift in community practice, but it seems to me likely that disciplinary pressures together with pressures within universities might help to explain it. There is certainly an issue of funding, its importance, availability and demands, which in many disciplines has led, when a research proposal is put together, to the establishment of teams that are broader than previously. Some research funders specifically mention multi-sites or multi-disciplines as criteria. This might be exercising influence in mathematics, also. A further consideration could be the role of the conference network, in providing meeting places, and the internet, in encouraging communication between community members. Many of the participants spoke of meeting up with their collaborators at conferences, or of meeting them initially in this way. It is possible that an oft-cited preference amongst females for working together has had some effect although the female presence in mathematics is not yet substantial enough to expect that it could provoke such an important cultural shift. And this over-simplistic stereotype of the collaborative female and the competitive male is, in any case, not borne out by my data.

Why did the mathematicians collaborate?

From the perspective of a mathematics educator, the grounds that the mathematicians gave for collaborating were very pleasing, reflecting as they did the reasons to be found in the general educational, and mathematics education, literature.

> *If you are collaborating with mathematicians, you feel much less isolated. Because mathematics is an isolating experience, you often feel burnt out, dried up, you have no skills and everyone else does. When you start working with someone else, you discover that there are things that you can do, perhaps better than they can, and things that they can do that you can't. You can draw on their knowledge without having to go through the pain of having to look everything up. It is like having a library that does its own selections for you* (Female Reader).

> *With a collaborator we make faster moves, because my stupid ideas get sat on faster. We also generate more ideas between us...I did some collaborative work about fifteen years ago where there were so many cases that I really wanted to give up. The colleague's tenacity paid off but, on my own, I probably would have given up* (Male Professor).

When the 66 mathematicians who claimed to be working with others, collaboratively or co-operatively, were asked why, they offered the following thirteen reasons, although, of course, any one individual would only mention a few of these. They are listed from the most, to the less frequently cited.

- Talking is a good way to get a problem done.
- It shares the work.
- You benefit from the experience of others.
- It increases the quantity and quality of ideas.
- You have someone off whom to bounce ideas.
- It enhances the range of skills.
- You get into areas that you might not have thought of going into.
- You learn a lot from more senior colleagues.
- Under the pressure of writing up, you must not let others down.
- There is someone to take over when you reach a dead-end.
- You share 'the euphoria' with someone.
- You feel less isolated.

- You can benefit from a novice/expert combination.

The world of mathematical problem-solving, presented here, is very different from that to be found in the mathematical literature where the results and the techniques are reified and the participatory practices of the mathematicians, as recorded here, are expunged. If, however, we think of these mathematicians as learners, their enquiry processes provide an excellent model to use for less sophisticated learners. Viewed as an educational process, then, and when based on participation as, in this case, a collaborative practice, Etienne Wenger pointed out that such processes:

> are effective in fostering learning not just because they are better pedagogical ideas, but more fundamentally because they are 'epistemologically correct', so to speak. There is a match between knowing and learning, between the nature of competence and the process by which it is acquired, shared, and extended.. (1998: 101/102)

I explore this further in Chapter 10.

Placing a classroom emphasis on collaboration, which requires discussion, and consequent sharing, might not only help to address some of the aspects of competition so disliked by some of the mathematicians, and by students, but would also be more likely to produce effective learning. After all, that is what the mathematicians are claiming. It is also what students call for. In 2001, I reported the results of a study in which students who had opted to take mathematics at 'A' Level (16-18 years) in secondary school, were asked whether they would continue with their studies. 73.3% unequivocally said they would not, including 84.6% of the females interviewed. Asked what kind of classroom they would like and to provide 'dos' and 'don'ts' for teachers:

> the students wanted their teachers to facilitate discussion, teamwork, a light-hearted approach, a relaxed classroom environment where you are not afraid of making errors. They told anecdotes to support the fact that they did **not** want to be put down, persistently asked the same question, made to look a fool or feel patronised, put into a position where others laugh at you, be thrown 'in at the deep end'. (Burton, 2001a: 67)

But collaboration was not all 'smooth sailing', as many of the mathematicians made clear when they spoke about competition.

Competition

One shift that might, but has not, taken place in mathematics is away from competition. The characteristics, which define, bound and direct the practices of mathematicians, I define as aspects of their mathematical culture. Within that culture, there appeared to be recognisable social practices, which were well known, sometimes mocked (see, for example, Davis & Hersh, 1983) and certainly not always applauded. Curiously, competition appeared to co-exist with collaboration, or co-operation, within the mathematics community, overtly sometimes in the same people. This was regretted as much by some of the men as by many of the women. Of these, some of the mathematicians spoke very negatively about the effects of competition on working practices, including two males:

> *In the States, I worked in an Institute where the work was unsupervised but the atmosphere was highly competitive. I became disillusioned with that, and the sense of self-belief that Americans seem to have but also the fact that everything had to be instantly commercially viable.*

> *Personally, I feel more comfortable dealing with things in a collaborative way. There can be tremendously savage competition in our field and that is not something I feel particularly comfortable with. If you work with people, they bring different things to the scheme of affairs. You can think more fruitfully that way because you have material offered by others to chew on. We hardly stop talking.*

Something similar was also heard from a female pure mathematician:

> *I have been struck recently by people's style at conferences. There is a particular culture which we are not used to where a seminar is a defence of your result against the audience which interrupts all the time, questions you aggressively, it is highly competitive. I hate that. I don't mind people asking questions. I like that. I like to be asked questions but I don't like that style.*

So, competition wove its destructive influence through the professional experiences of many of the mathematicians, in seminars as well as occasionally in the inter-personal exchanges within a collaboration. A female commented on:

the collaboration [which] was extremely difficult [...] and I think that was because we were treading on each other's toes[...] It was also too competitive and the wrong things were made too important. One of the collaborators was used to a very argumentative style and I hated it.

There were those who valued its effects, but many others who decried them. Although resisted by a number of my participants, especially but not solely the females, the competitive spirit appeared not only to be pervasive, but to define and distort the experiences people had, again especially for the women. As already stated, competition featured in respect of working practices which were related, in some instances, to too few departments in England in a particular speciality and to being in a fast-moving field. One female pure mathematician, speaking about writing practices said:

I think that scholarly articles are often written to impress people how clever you are. I try not to do that. I don't like it. If you explain something well it can sound quite simple but you don't want people to think it was easy, because it wasn't. The same thing applies to seminars. So often they can put up hundreds of equations that you can't take in, you might have no idea what the person is talking about after the first five minutes. I think this is a waste of time.

Other participants emphasised how important it was to be clear and accessible when writing, or speaking:

Mathematical quality alone is not enough. I am very conscious of how the ideas are expressed. I like a story - a paper, a seminar are both stories. There is a beginning, a middle and an end. So, well-written, clearly explained in a manner which is easy to follow and logical.

But, the way in which power, and consequently competition, 'normalises' other practices, for example journal editing, was also made clear: *You learn that you certainly do have to write all the letters in exactly the form the editor wants or else you won't get to referee those papers and they won't referee yours.* In Chapter 8 this issue re-surfaces in a discussion about publication.

Mary Daly (1997), drawing on the work of Lorraine Code and Helen Longino, pointed to the mutual construction of knowledge *and* the knower within their dialogic communities. She said:

> It is a vital aspect of the values of objectivity and rationality, emotionality, intuition and subjectivity, that the composition of the knowing communities is critically analysed in order to see how the community membership and its practices and power relations inform the knowledge created, refined and acknowledged. (p.49)

In the case of the mathematicians whom I interviewed, such an acknowledgement, never mind an analysis, was not present. A number of them were critical of the ways in which competitive and hierarchical practices impinged upon them, as we heard above: *It was too competitive and the wrong things were made too important.* But such practices were not related to the values that often they themselves accorded to 'objectivity' and rationality.

Where and when is competition first encountered? A female mathematician drew attention to the influence of school and environment on the formation of competitive habits: *I am competitive and obsessive and I am not sure I should be encouraged in that. The school was like that, my background is like that* whereas a male mathematician remarked uncontentiously about his secondary school: *there was a competitive thing as well.* Another participant speaking about her undergraduate studies said: *The thing I hated about doing mathematics was that the boys were so competitive.* Jo Boaler (1997a) discussed the reasons given by schoolteachers of mathematics, especially those with setted (tracked) classes of high achievers, for why they chose to pursue a teaching strategy of pace and pressure. They have no evidence to support their position and it is not acceptable to many students.

> One of the reasons commonly given for the formation of setted groups is that the competition created by setted classes helps to raise achievement [...] However, of the 24 students interviewed in Year 11, only one student, Gary, gave any indication that the competition and pressure created by their setted environments enhanced motivation or learning. (p.128)

Kathleen Lynch and Anne Lodge (2002) observed "patterns of dominance" in secondary school classrooms, particularly those of mathematics, "including the competitive seeking of power and control within the classroom" (p.129). It is my contention that this helps to define the mathematical culture into which learners are embedded and which becomes part, for some, of their history of mathematical learning success. The mathematicians reported here have made it clear that this is not the case for all. But it appears to be dominant in the culture and is something about which more is said in Chapter 10.

I am suggesting that, in these subtle ways, power and consequently social positioning are imposed by means of the conventions of those practising the discipline. They are established and maintained through the social practices, for example with respect to writing and publishing (see Chapter 8), but also in the rules by which the game of mathematics is played, especially interpersonally (and see Chapter 9 for a further development of this argument). And most of all they are embedded in the norms acquired by new entrants to the profession as part of their socialisation as mathematicians.

SUMMARY

The participants in the study did not work on only one problem. They had ways of keeping a number of problems 'on the boil' such that no progress on one allowed them to shift to another. These problems were sourced from people, at conferences, seminars, the internet or from text. But whether the collaborator or the problem came first was a chicken and egg story. For example: *I tend to find problems that interest me although some people do come and find me.*

Sometimes resolution of a problem was more a question of a happy serendipity than meticulous application, such as in the story told by one participant who had worked on and off for several years on a theorem where *the proof didn't work but I knew the theorem was true.* Some years later, at a seminar, the presenter

> *was talking about something that was somewhat relevant and he got it wrong [...] What he was talking about was a simple case of what I was trying to generalise. Twelve hours later, I got on the plane, got to Australia, got up in the middle of the night and wrote it all out and it was perfect. It wasn't an idea that was missing, it was the notation.*

With the identification of a problem, the mathematicians also identified collaborators such that, contrary to my expectations, 66 out of the 70 worked either co-operatively, across disciplines or areas, or collaboratively with other mathematicians. The reasons that they gave for so doing matched those given in the educational literature for encouraging learners to work together. However, interpersonal collaboration co-existed with disciplinary competition and many of the mathematicians described competitive practices, that some regretted, between colleagues, in seminars, at conferences, in publishing. The competitive norms they described sounded very much like the norms of the exercise of power, about which more is to be found in Chapter 9. The pedagogic implications of collaboration and competition reflect competing epistemologies. Collaboration implies team or group working, interpersonal respect, the construction of knowing in learning communities. Competition, on the other hand, reinforces individualism, hierarchies, elitism, positioning and the transmission of 'objective' knowledge. These implications, as well as those of other practices outlined by the mathematicians, are pursued in Chapter 10.

So, the epistemological heterogeneity was supported by very interactive, and collaborative working practices even where the mathematicians themselves referred to an overall competitive spirit within the discipline itself. Knowledge sharing was spoken of in very positive terms. But knowledge dispersal, in mathematics, invokes the traditional journal publishing routes which themselves carry particular community constraints. These are discussed in the next chapter, Chapter 8, where I pursue the issue of publication.

8

Writing for Publication

One mathematician remarked: *I get very annoyed at having to write things down on paper. It's slow, cumbersome, you can't do it anywhere.* Nonetheless, publishing is a necessary part of academic life in universities, even more so in the UK since the introduction of research assessment exercises. What 'should' be published? Why? How do mathematicians learn to maximize the chances of publication of their own work? What criteria do they use for judging the work of their colleagues? The answers to such questions constitute an important part of the mathematical culture and of the culture of mathematics. They emphasise, yet again, the heterogeneity of the mathematicians who, while sharing a culture, adopted very different perspectives within it.

During that part of the interview when the mathematicians had been describing how they thought about the research problems on which they worked, I said something like the following:

> Leone: *So someone sends you a paper from a journal and asks you to review it, or you send to a journal the paper that you have written of the piece of work that you have been describing. What are the criteria that you bring to the judgments that you make and/or that you would hope that the reviewers of your paper would bring?*

In their responses, there was a widespread demand for 'rigour' associated, as with so much in the world of mathematics, with hierarchy. This consisted, in the main, of the application of an hierarchy of judgment which started from significant, decreased in value to important, then interesting and, finally, trivial. (See Chapter 9 for a discussion of the impact of hierarchy on the mathematical culture.) One participant said: *Interesting is a polite way of saying it hasn't done very much. Significant means that it has made a major contribution whereas important indicates a contribution.*

Surprisingly, only 19 of the mathematicians demanded that the work be correct and fewer saw it as their responsibility as a reviewer to ensure correctness although not mentioning correctness does not mean that it was assumed or felt to be unimportant. A few more, 28, required something new, mentioning originality or novelty. Inevitably other undefined terms, such as elegance and beauty (discussed in Chapter 5), were introduced, sometimes to attempt to describe or position the value judgments implicit in words like significance. In this chapter, I describe these responses, first to rigour, then to the hierarchy of judgment. I continue with a discussion of correctness and novelty and conclude with a review of three issues raised by a number of the mathematicians, writing clarity, length and connections. (The implications for the writing of students, as well as teachers, are explored by Candia Morgan (1998).) I begin with rigour because of its importance to, especially pure, mathematicians but also because it is so difficult to define.

RIGOUR

> *What I found so exciting was the rigour. It is the logic of the argument with as few basic axioms as possible and as elegant as possible.*

This mathematician provided the beginning of a description of rigour in logic of argument, succinctness and elegance. Philip Davis and Reuben Hersh (1983) found rigour more elusive. In writing of The Ideal Mathematician, they said:

> He rests his faith on rigorous proof; he believes that the difference between a correct proof and an incorrect one is an unmistakable and decisive difference...Yes, he is able to give no coherent explanation of what is meant by rigor, or what is required to make a proof rigorous. In his own work, the line between complete and incomplete proof is always somewhat fuzzy, and often controversial. (p.34)

In Chapter 5, I quoted a comment from Hadamard that rigour has no other object than "to sanction and legitimate the conquests of intuition" (quoted in Howson, 1973: 78).

For most of the mathematicians to whom I spoke *rigour is important* but for many *it doesn't turn me on particularly.* The differences between pure and applied mathematicians surfaced in this discussion too. One applied mathematician said: *I think mathematics as applied in engineering should be explained whereas mathematics taught for mathematics' sake should be rigorous.* Below, I quote extensively from the transcripts as what the mathematicians say demonstrates a number of features to which I have been pointing in this book, features such as the distinctions drawn between pure and applied mathematics, competition, put-down, and justification. The next two applied mathematicians put different perspectives:

I don't think of rigour in the same sense as pure mathematics. I don't identify rigour with those techniques. I have a slightly ideological position which is that the problem with some pure mathematical techniques is that the techniques seem to displace the ideas even in the minds of the pure mathematicians and I don't see that as a tremendously strong intellectual position. There is a flavour about certain aspects of pure mathematics which I don't find very appealing. Of course pure mathematicians have a very different view. Linear algebra, for example - it is a tremendously powerful tool and a rationalisation of all sorts of ideas. But if you take what is called analysis, that still seems to me to be a ragbag of techniques that never gets crystallised into a coherent intellectual structure. In applications, rigour is often inappropriate because, even if you haven't done things "rigorously" (in the pure mathematician's sense), you will have thought it out and you will have covered various issues, perhaps thinking intuitively or perhaps thinking from a physics point of view. So it isn't as if you are approaching it in an empty-headed way but the detailed rigour is often not very interesting.

Physics unfortunately has the word approximation built into it as part of its intellectual vocabulary. A hard-core mathematician is likely to say that things that are approximations, or not always true, are not rigorous. Mathematics is either right or wrong. So if someone feels my mathematics is not rigorous, I will get upset with them because my mathematics is either right or wrong. If they don't like it, they should do better! For example, some analysts say functional integration isn't rigorously defined and we don't want to believe anything that comes out of it. But you can have very interesting results in topology from it. I think it is an abuse of the concept of rigour. You do get results and they are true results. There is a particular kind of English frame of mind that uses rigour like that. In the States, they will ask can you do any better but they won't accuse you of being unrigorous or wrong. People here, for reasons I cannot understand, tend to reject those kinds of results.

So while most of the mathematicians agreed on the importance of rigour, many, particularly applied mathematicians, dissociated themselves from its application to *their* work. Or they used 'rigour' as a stick with which to beat pure mathematics. However, in doing so, as well as putting a different slant on the reliability of pure mathematical techniques, they sometimes used somewhat tendentious arguments! This is supportive of the argument made by Wenger (1998) in which he stated that the engagement in practices by those within a community is not in one-to-one correspondence with the reified structures of, in this case, the discipline of mathematics (see pps118-121). If one follows this line, within the discipline there is a 'public' view on mathematics, which includes notions like rigour, and a 'private', or what

Wenger called an "informal" practice (1998:118) that serves the purposes of a more restricted community, such as that of applied mathematicians. The distinctions between public and private rarely surface except in situations such as these interviews provided.

Another mathematician said *I don't like things that aren't rigorous. No sloppy loose ends* but admitted a short while later that there are some things that must be taken at face value, no matter how rigorous you think your argument is: *even though, when you write out what you think is a rigorous argument, at some level you have to accept.* So, rigour is not necessarily all-encompassing as unstated axioms may underpin an argument for which rigour is apparently claimed. *But there is an expectation that mathematical arguments should be presented in such a way and to a level of rigour that enables the reader to feel sure that the results are correct but without having to present every detail and turn over every stone along the way.* This "level" of rigour appears to be subjective and can lead to the overuse of such words as 'clearly' and 'obviously' (and see the discussion in Burton & Morgan, 2000, Appendix C). A mathematician acknowledged this:

> *Rigour is avoiding hand-waving, proving every argument in detail, 'it's obvious' - that is not acceptable. I may write it but when I prove a result I avoid it and I don't like using it. It is important to leave out any possibility of not having proved everything in detail. So 'obvious' leads to mistakes. Rigour is adding understanding because of the detail of the proof and also when you prove something non-rigorously you have the impression that you have proved it but you haven't really. So essentially avoiding rigour is avoiding some bits of which you are uncertain.*

And another mathematician warned:

> *When something is obvious to me, you have to be aware of the pitfall of not dismissing something, but you have to be pretty sure that you can go away and prove it even if it takes the next ten years.*

By no means everyone was happy with rigour: *I am a little ambivalent towards rigour in that it is certainly necessary to attend to the quality of maths but people tend to concentrate on rigour at the expense of explanation.* Someone else was more definite:

> *I don't like rigour. Rigour to me in the first instance implies time-consuming non-necessity. For a mathematician, that probably sounds terrible. If you apply rigour to proofs, then that is where you get the beauty again. But I don't want to have to apply the rigour. I want somebody else to apply the rigour and me to see it. I think it is tedious. I know that in mathematics rigour is essential but maybe going back to the way it was*

> *taught. I don't like that sort of tedium. My problem as a statistician is that the obvious is generally not true. So I cannot say to myself that is obvious and therefore I know it and therefore why should I bother with this rigorous proof because I know that the rigorous proof is essential. It is not my forte. I like to solve a problem and it might take rigour to get through it but I wouldn't like to be doing it for days and weeks. I think the rigour is important - it is paying attention to detail and I certainly think you must pay attention to detail. The beauty of mathematics is that you can take account of all eventualities that you have thought of and that is rigour. There is no point in applying a theory that you don't know will work in all situations. You have to know the situations in which it will not work.*

But concluded by saying: *But I am not a pure mathematician.*

So like many of the descriptors used by the mathematicians, rigour is elusive. It means different things to those in pure, applied or statistics. The judgments it entails are subjective. Not everyone can be bothered with its demands. And, in any case, even a rigorous argument cannot be completely bounded. But it is maintained to be of central importance!

> *I am not sure that I understand the difference between correctness and rigour. Some things are true in smaller universes which can only be true in larger universes if you rigorously follow an argument.*

> *Rigour is the ability to re-construct the argument. But of course, with different people, different amounts of detail are adequate.*

> *That is what rigour is - the ability to convince someone else of truth. I am demanding that a paper convinces me that it is true.*

So what *is* rigour?

> *I think rigour is equally subjective. Rigour is something that you learn to use appropriately. A standard of rigour depends on the context. You tend to develop ideas of what is appropriate to a given context. You could not say that there is a uniform standard of rigour.*

It seems clear that rigour is a social construction and that agreement about it is culturally construed. The next mathematician began to define the boundaries around mathematics and to agree closely with Wenger's treatment of boundaries (see 1998: Chapter 4). In the process, agreement was also expressed with the social nature of the discipline and its practices:

> *There is a social thing, a very clear idea of what a correct proof is and that is rigour as I understand it. Something isn't part of mathematics until proved that way. There are parts of mathematics that cannot be proved in*

that way, particularly things connected with physics, where there are partial results but that is often the most interesting stage...

I don't think mathematicians have any real disagreement about what a rigorous proof should be. Of course some proofs are not rigorous because the writers don't have the necessary expertise but, more interesting, sometimes we just don't have the mathematical structures needed to turn the heuristic arguments into rigorous proof. But you could argue that the notion of what is a rigorous proof is a product of the society of mathematicians but I don't think there is any disagreement about what it is.

While it may be the case that there is 'no disagreement' about the recognition of a rigorous argument, disputes about recognition can be the substance of articles, books and reports. Brian Rotman explained it in the following way:

> There is nothing intrinsically wrong with or undesirable about 'rigor' in mathematics. Far from it: without rigor, mathematics would vanish; the question is how one interprets its scope and purpose. On the present account, rigor is not an externally enforced program of foundational hygiene, but rather an intrinsic and inescapable demand proceeding from writing: it lies within the rules, conventions, dictates, protocols, and such that control mathematical imagination and transform mathematical intuition into an intersubjective writing/thinking practice [...] Mathematics is not a building – an edifice whose truth and certainty are guaranteed by an ultimate and unshakable support – but a process: an ongoing, open-ended, highly controlled, and specific form of written intersubjectivity. (2000: 58-59)

Disputes focus on whether or not a proof does prove and consequently on the intricacies and complexities of the proof argument, and the judgment criteria used. However, the papers published in the journals are all peer-reviewed and the mathematicians were quick to invoke a hierarchy of judgment for this purpose. Significance was the most valued, followed by judgments of important, interesting or, finally, trivial work.

HIERARCHY OF JUDGMENT

Significant contributions are what should be in leading journals whereas interesting contributions often fall short of significance. The criteria for making these kinds of judgments are passed from academic father to academic son, if you like.

There are a number of interesting, cultural assumptions in this statement. The mathematician appears to agree with the notion that 'everyone knows' what is significant, that criteria 'exist' and it is possible to apply them. This cultural knowledge is acquired through the usual powerful, masculinist, osmotic, forms with the underlying assumption of the non-presence, lack of appropriateness, of the female. (And see Chapter 9.)

Significant

How is a significant paper recognised?

> *Significant results are of relevance to other problems, looking at things in a way that people haven't thought about, applying some method to a problem which hasn't been tried before. Important and significant I would put fairly close together.*

But the distinction between significant and interesting, for the next mathematician, who did not mention 'important', was clear:

> *A significant problem leads to something else. It is a breakthrough on a longer trajectory. Tidying some things up might be interesting. It can be a cute result. Interesting is already something that has some novelty, perhaps some application, perhaps illustrates some nice techniques. But it is not the breakthrough on the path to somewhere. To be significant it has to be the paper which is the turning point in people's understanding of some subject.*

Important

The relevance to significance of achieving this breakthrough is interesting in the light of the next comment that moves us to important.

> *It is hard to spot important papers at the time because quite often an important paper will not just be more of the same but breaking new ground. Most people won't really understand what it is about or recognise the new ground until later. I don't use importance as a major criterion. If there is potentiality, that is good enough.*

The space between significant and important appeared to be very narrow:

Important is much the same as significant. It has to be really different on what's gone before. It has given you something substantial. It doesn't merit publication if the result is boring.

Interesting

Just how fluid were the categories is demonstrated in the next quote:

Interesting - it might be brand new but it isn't significant if it is local. Something that I think might get cited by other people is interesting; it moves the subject forward in some reasonably significant kind of a way. Interesting is obviously subjective to some extent and one tends to be interested in the things that one is working on oneself.

I gained the impression that applied mathematicians regarded 'interesting' somewhat differently from pure mathematicians. Some regarded an interesting paper as one that had something to interest them:

An interesting problem is something that has a physical reality associated with some mathematics which is non-trivial. Interest is probably something which is very close to what one is doing oneself.

Whereas, others seemed to treat 'interesting' as a judgmental criterion:

At one stage I did decide that every paper I published should have at least one interesting idea. I have never published a paper just for the sake of it. It has always had something slightly non-trivial. Which isn't true for all papers I have ever read.

And with some it was hard to tell:

Interesting is regarded more seriously by some journals than others. Journals where there are communications in mathematical physics and nuclear physics demand more of interest than those in classical and quantum gravity. Interesting means a result that is deep rather than anyone could have done this. It either means hard work or something particularly clever.

However, there was no disagreement about the category of dismissal, trivial.

Trivial

I am quite a harsh referee. I wish I didn't reject so many. I like to think that I just don't get the good ones. I like to see a result which is non-trivial (another subjective category). Trivial, despite being *"subjective"*, was freely invoked to make negative comment: *Trivial - they aren't really saying anything different, it is a re-hash that might have been said before, and possibly better.* But not necessarily only of the other: *Typically of mathematicians, you prove a result and once you have proved it you understand it completely and to you it seems trivial. You prove something and suddenly it's obvious.*

CORRECTNESS

Should a submitted paper contain correct mathematics? There was no agreement. One male mathematician had a somewhat flexible view on correctness:

> *So the paper has to be entertaining to read, non-trivial, a substantial result, clear and easy to read except maybe the proof. Easy to read is very hard to do. By and large, depending on the journal, the thing has to be incorrect in a substantial way to be rejected.*

How did correctness feature in the judgments that were made? Did the mathematicians check every paper they were sent to review to ensure it was correct? I had experiences to suggest that this was not likely as some papers in the journals do contain mathematical errors.

Much to my surprise, although many of the mathematicians felt that the paper should be correct, they did not see it as the job of the referee to check whether or not it was. Few reported doing the mathematical work to ensure correctness in others' papers and not even always in their own. An exception said:

> *Sometimes the feeling of completion and correctness misleads you and you see that it is blatantly incorrect when you return to it, as an outsider. Or maybe you submit a paper to a journal and you come back and look at it a month later, or someone else from outside maybe refereeing your paper, gives an indication.*

Although many said something like: *I require of an article that the mathematics is correct,* it was not clear if this 'requirement' had to be met by the author or was checked by the referee or the journal editor: *By and large, depending upon the journal, the thing has to be incorrect in a substantial way to be rejected.* But some mathematicians provided more detail:

> *Mostly I assume correctness. When I started I would work through the detail. Now I don't have the time to do it and sometimes the amount of detail is so immense that you cannot work through it.*
>
> *Is it correct? which is easy to answer but I don't always check the mathematics because it is dull to do that.*
>
> *I don't interpret a referee as having to be responsible for correctness unless they are particularly asked to. I don't think a referee's job is to guarantee correctness. On the whole, if a knowledgeable referee has a positive overall impression then any mistakes could be corrected i.e. are not fundamental flaws in the argument.*

However, as with everything else, there were those who took the opposite position:

> *I would expect it to be correct and I would check a paper that I was sent. I expect it to be written to a standard. To be organised and structured and properly referenced.*
>
> *I would rather expect the referee to check that it is actually correct as a minimum and I am disappointed when referees don't do that for mine. Often they don't.*

Given the evidence for poor written mathematical communication (see, for example, Burton & Morgan, 2000 reprinted as Appendix C in this book, Hersh, 1989) and the advice offered by professional organisations (e.g. Gillman, 1987, Knuth et al., 1989), the charges against philosophical writing represented in the following comment of a male mathematician are interesting:

> *a philosophy paper which employed mathematical notation to obfuscate and hide the absence of genuine thought. I would contrast this with notation in mathematics which almost always serves a useful purpose.*

One mathematician had a list of requirements: *Originality, being theoretically correct (mathematically and physically), interesting to a broad readership are all important for publication.* Originality and/or novelty were discussed by many.

NOVELTY

New is essential and it means that I have not seen it before.

I would like to think that my papers have a novel feature, something that hasn't appeared before theoretically, perhaps a new mechanism for understanding something.

I would want a published paper to offer original work, something that we didn't know before, interesting work not some trivial thing that maybe is new but who cares?

The mathematicians quoted above were like many of the twenty-eight who stated an expectation of novelty or originality in published papers. I do not believe that, in this respect, mathematicians are very different from those in other disciplines. However, what strikes me as different is that the novelty was rarely explained as, in any sense, connecting to known results, except in the following two cases:

I would look for something substantially new in a paper. It should make an original contribution to the field. I would include new results, possibly extension of previous work but new results for which we didn't have physical data. It may not be a totally new model but it would be applying an old model to new situations or a new model. So part of it would be original in the sense of new data or new method.

It should either be something new or a new derivation of a known result. It should be made clear how this new work, or this new derivation fits into what has been done already.

The next mathematician offered a summary of issues that were taken up in a number of the interviews and have been discussed:

Criteria - interest, correctness, conciseness. Is it new? Is it interesting? You can tell what newness is from your past experience. Then you have to work out if it is actually right. How is it written? The important things are whether it is doing something different from what other people have done and if it isn't doing that, it probably shouldn't be accepted.

ISSUES – WRITING CLARITY, LENGTH, CONNECTION

Writing clarity

Twenty-eight of the 70 mathematicians, 12 females and 16 males, made comments about clarity of writing. One mathematician summed this up by saying: *the communication problem in science is pretty hard.* There were many references to poor writing expressed as obscurantism, trying to impress with your cleverness, poor referencing. One participant said: *it must be written clearly enough that you can get the information out of it without killing yourself in the process.* Another said: *I am not keen on badly written stuff. If it is moderately interesting and badly written I will turn it down.* This was not the only mathematician to be critical of writing style. For example, *notation and things like that shouldn't be a problem.*

Many of the mathematicians explained their own approach. For example: *I think one should be open about the limitations of what one's done, the problems and possibilities.*

> *So in writing, I try to give the structural idea before getting into the nitty gritty detail of what you have done. If I do a good introduction, I kind of give the conclusion in the introduction and I would point the way towards what might be done. I would include references to the applicability of what I had done.*

> *I would like my papers to be understood and well written and I would like the ideas to come across. The basic ideas to be clear, the insights, the understanding and the resolution to come across.*

> *When I write anything, I think of the things that I have suffered when people do not explain things well. I think things should be set out very clearly from the beginning. Structure, design, well organised, the methods used should be explained. The abstract is important in medical statistics, for example; the doctors won't read the whole paper [...]*
> *Everything should flow easily - shouldn't give you a headache or a tummy ache going from one line to another.*

Some made cynical remarks about the process:

> *I don't believe the refereeing process for publications is independent of the referee's perception of the paper. So if their background is in your area they might say it is full of trivial details and if it isn't they may say it doesn't have enough detail.*

> *It is also quite important what name you have in the author list. Where there is a famous name in the list of authors, the paper is much more likely to be accepted. This is a bit annoying but justified since this rather famous person must have some good reason for publishing in this way.*
>
> *In mathematics if somebody who doesn't really know an area referees a paper they will not necessarily recognise its value or lack of value.*

I imagine all would agree with the statement made by one: *I would like to be made to feel WOW! That is really pretty.*

Length

Thirteen mathematicians commented on the length of papers: *It shouldn't be too long. I have sent people's papers back for re-writing.* This is a little like the argument about how long is a piece of string. The paper should not be too long, but another mathematician pointed out, it should not be inappropriately too short either: *It should not be able to be solved in one or two lines using more advanced techniques than are offered in the paper - sophistication that is suitable to the problem.*

Brevity or conciseness was a criterion that was mentioned by a number of mathematicians and, for some, seemed particularly related to elegance or beauty. The concerns of two mathematicians surface below:

> *The problem that we worried about was whether people would think that the length justified the excitement of the result. However nicely you have done it, the end product has to be of sufficient interest to justify the tortuous argument in getting there; or can you develop a sufficiently elegant argument for the argument itself to be interesting. Elegance is very important. Having solved it tortuously, the next problem is to solve it elegantly. It is all very relative of course.*
>
> *I do think there is a difference between a slick and a non-slick proof but brevity is often preferable to length for purely utilitarian reasons. In a hundred-page proof you might develop some interesting ideas where a short proof might be opaque. One of the problems about some subject areas in mathematics, is that they develop powerful machinery which is hard to understand. It may be easier to solve a problem with it but you don't understand what is going on. There might be more beauty in the details of a long proof than in a short terse proof that doesn't tell you very much.*

Connection

Twenty-four mathematicians spoke specifically of the value of connections that might be set up in a paper between this paper and published work by others, between different areas of mathematics, between theory and application. It might be that *It relates two apparently unrelated topics.* Or *these two things that I thought were unconnected are actually the same thing. It rounds your picture off in some ways.* However, in the interviews as a whole, there was universal agreement about the importance of connectivities (see the discussion in Chapter 6) so it is not surprising that the making of connections was something that just over one-third of the participants spoke about when discussing publication.

In a paper written with Candia Morgan (2000), we presented the results of a discursive analysis of 53 published papers obtained from the participants in my study (reprinted in this volume as Appendix C). We found close links between their epistemological positions and features of their writing style, and connections between philosophy and presentational style, which are also noticeable in texts. We said:

> 'Common knowledge' about the nature and process of mathematical writing (as represented, for example, in the MAA guidance mentioned above) suggests that there is only one standard way of presenting one's research. Yet our examination of this relatively limited sample of published papers has shown considerable variation and, in a few cases, wholesale flouting of the recognised conventions...Knowing the conventions and being able to use them may be one step towards establishing one's position; knowing how they work and how and when to break them in order to achieve a particular effect is, however, an important way to express and establish a more powerful position. (p.450)

So, while the mathematicians adopted very clearly expressed positions on what should be published and why, these positions were rarely supported by any serious analysis of the relationship between what they were writing and how. The result appears to be that publication becomes a bit of a lottery but not an unbiased lottery since where you are in the hierarchy, what is your speciality, with whom you are writing, and how that writing is presented are all influential on whether or not the paper is actually published.

SUMMARY

I began this chapter with the following four questions:

What 'should' be published?
Why?
How do mathematicians learn to maximise the chances of publication of their own work?
What criteria do they use for judging the work of their colleagues?

While the mathematicians appeared to be clear about what should be published and why, when this clarity was explored it appeared to rest on socio-cultural practices interpreted differently by different individuals. Learning what these are, and how to use them effectively to gain access to publication, was claimed to be a product of luck, knowing the 'right' people, experience but not deliberation. Wenger (1998), on the other hand, would explain these socio-cultural practices as arising from the community boundary practices, those that the insiders learn to apply, and those that are intended to exclude the outsider. He said: "As communities of practice differentiate themselves and also interlock with each other, they constitute a complex social landscape of shared practices, boundaries, peripheries, overlaps, connections, and encounters" (Wenger, 1998: 118). There were those who expressed concerns about, or at least recognition of, some of these practices but there were also many who were happy to make judgments on bases that were extremely slippery. This was particularly noticeable in the discussions of rigour but equally so in the hierarchy of judgment that was invoked by almost all of the mathematicians.

When it comes to criteria for publication, one would imagine that correctness, in mathematics, would be of supreme importance as, indeed, it was. This made it all the more surprising that so many of the mathematicians did not think it was the referee's responsibility to ensure correctness of a submitted paper and very few articulated strategies for ensuring correctness of their own work. In some cases, there seemed to be a curious non-circularity where the author of a paper relied on referees to ensure correctness, but as referees, they did not necessarily see that as their job. One mathematician also pointed out: *It could be correct and clear but is it worthwhile?* This seemed to me to be an extremely serious question that does not necessarily get asked as part of the publication process although the application of the judgmental hierarchy, significant, important, interesting, trivial, is the way in which, implicitly, the mathematicians believe they are dealing with worthwhileness.

Novelty, or originality, was another criterion that featured in the discourse of about 42% of the mathematicians. But the assumptions that lie behind novelty, meaning 'new to me', were not exposed. Further criteria that surfaced were writing clarity, length and connectedness but in every case there was little substantial agreement about what was meant.

In the paper by Candia Morgan and myself, referred to above, we demonstrated both:

> [...]diversity in the variables we have examined and the lack of universal conformity to commonly acknowledged conventions. Other strong influences on writing-style choices may include the authors' epistemological stances on mathematics as well as their beliefs about the expectations of their readers. Nonetheless, our analyses of the various aspects of identity demonstrate that a research paper does more than report research. We have shown how authors, through their writing, convey very particular perspectives on themselves, as well as on their mathematics, and on how they build their identities with respect to their communities. (2000: 449)

The importance of writing, to a subject like mathematics where communicating through publications is a professional necessity, is clear. But writing is also central to activities in classrooms; there are, consequently, extensive implications for teaching and learning mathematics, particularly in relation to the lack of specific access to information about the impact of writing and how to use its forms to tell your own story. These are taken up in Chapter 10. But there are also issues about gatekeepers and the exercise of power that are addressed in the next chapter which presents the differences between female and male mathematicians in their experiences of the mathematical culture.

9

Strangers in Paradise? The construction of mathematics as a male community of practice

> One of the reasons why there appears to be no alternative to our mathematics is because we routinely disallow it. (Bloor, 1991, 180)

The study that I have been reporting in this book was, in part, attempting to answer two questions: Is the academic community of practice of mathematics recognisably defined? How might it look and feel different? In particular, I was concerned to know if women's experiences within the mathematical community and their ways of coming to know mathematics were distinctive and, if so, in what ways. The term 'community of practice' was described in Wenger (1998). For Etienne Wenger, "living is a constant process of *negotiation of meaning*" (Ibid: 53, original italics), the purpose of which is to learn. The negotiation of meaning involves "the interaction of two constituent processes...*participation* and *reification*" (Ibid 52, original italics) and "their complementarity reflects the inherent duality of this process" (Ibid: 66). He described their relationship as "on the one hand, an intense involvement with the reificative formalisms of [the] discipline; and on the other, a deep participative intuition of what those formalisms are about" (Ibid: 67). He pointed out that:

> it is the interplay of participation and reification that makes people and things what they are. In this interplay, our experience and our world shape each other through a reciprocal relation that goes to the very essence of who we are...the river only carves and the mountain only guides, yet in their interaction, the carving becomes the guiding and the guiding becomes the carving. (Ibid: 70/71)

Knowledge was described as "a matter of competence with respect to valued enterprises" and knowing as "participating in the pursuit of such enterprises" (Ibid: 4)

My concerns were to focus on how mathematicians understood their own knowing and the process of constructing it, which I understand as learning. Hence, participation and reification were fundamental to my analysis. They contribute to the duality of the social and the individual represented in the practices of a community as well as to the formation of and influences on its members' identities. So, far from coming to know being objectively describable or predictable, I wanted to explore this process as essentially human, and consequently socio-cultural. Discourse is central to issues of enculturation. Trevor Gale and Kathleen Densmore quoted "Stephen Ball's (1994: 21) observation that discourse determines not only 'what can be said, and thought, but also ...who can speak, when, where and with what authority' " (2000:64). Although it is not the only aspect, because of its importance it is discourse on which I now focus. (For a useful similar approach to scientists' discourse, see Gilbert & Mulkay, 1984.)

I began the study with the expectation that the process of constructing a research mathematician would involve a socio-cultural induction into both the mathematical culture (that is, the socially organised practices) and the culture of mathematics (that is, the disciplinary practices). For a mathematician who had entered that community of practice, I expected this process to be so robust that gender would not feature as a distinguishing characteristic. That is, my expectation, which was fulfilled, was that the range of practices, beliefs and attitudes about coming to know mathematics would be represented across women and men irrespective of gender. However, although the categories of the model were freely invoked by these mathematicians (see Chapter 6) their discourse made apparent that, through their social positioning, the women never quite managed to be taken, or in some cases to take themselves, as equivalently entitled to engage in all the practices of the community. So, the heterogeneous ways that women and men came to know mathematics did not appear to be gender differentiated; but their experiences within the world of mathematics, that I have called the mathematical culture, were distinctive both in terms of how they were treated by their colleagues, and how they approached their research and associated practices (for example, their approach to supervision both as, and of, a research student, see Burton, 1999a and Chapter 10).

In her novel, *The Mind-Body Problem,* Rebecca Goldstein (1993) presented Ava, a physicist, who said:

> Look around at the women in academia, the women who make their living from their brains – especially those in the so-called masculine disciplines like math and physics, to take two random examples. They all feel it too. They're telling you with the way they look and dress, the way they hold

> themselves and speak: feminine is dumb. You've got to stamp out all traces of girlishness, if you want to be taken seriously. (Goldstein, 1993, 194)

In this chapter, we hear how the women mathematicians and their male colleagues, whom I interviewed, described their different ways of experiencing and accommodating the social positioning to which they were subjected. I begin by discussing the community of practice of mathematics before exploring the discourse of power revealed in the interviews. Power was located in three areas, becoming and being a mathematician, community practices and publishing. Compared with the comments of the women, very few of the 35 men volunteered remarks under each of the sections discussed. Overall, for every male such comment, there were three or four female comments.

THE COMMUNITY OF PRACTICE OF MATHEMATICS

The mathematicians defined their community both well, and fuzzily, but they were clear about their choice of community. Some introduced boundaries: *I am not a mathematician, I am a statistician* or *I am a numerical probabilist* while others maintained flexibility: *I am a mathematician, a computer scientist...I don't identify with one particular community.*

Such heterogeneity of position, reflected in the choices they made about how they described themselves, was independent of gender and, in Etienne Wenger's terms, contributed to their identities as mathematicians as they negotiated the meaning of their experiences and related these meanings to their community practices and their knowings. Participation and reification were central to the functioning of this process. So, in becoming a statistician or a numerical probabilist, the mathematician learns how to *be* in such a community, what are its boundary conditions and what are the acceptable and unacceptable, often unstated, rules of behaviour, whether mathematical or social.

What are the practices that define the community of mathematicians? According to the responses of my research participants, many of these practices could be described in terms of aspects of the use of power (see Foucault, 1980). As one male mathematician said: *A lot of things are community based; if you are not in that community you will be unknown and looked at with suspicion* and competition operates across countries as well, sometimes inducing behaviours which some find unacceptable, such as the *one-upmanship and stuff like that* mentioned by the mathematician in Chapter 7 who went on to say: *It came very heavily from the French school!* A woman, who spoke of people ranking each other, continued: *Italian algebraists only do little problems* writing off the whole of Italian algebra at one go.

I encountered the discourse of power at 3 points in the interview. These were when the participants were discussing learning and teaching as part of their professional development and current commitment, when they talked about practices such as seminars and when they talked about publishing. I deal with each of these in turn.

Discourse of power

Becoming and being a mathematician

By the time the female mathematicians had entered the world of research, many of them had also recognised, and sometimes accepted, their individual positioning. However, understanding this as an individual experience is to mistake the nature of the functioning of the community. Power, fashion and 'intellectual snobbery' induce discipline-based behaviours. Here are the voices of three women, followed by three men. The women spoke about the need for protection, the 'specialness' and, perhaps, the resultant intellectual snobbery:

> *I have pursued a branch of theoretical statistics which very few people in Britain are doing and we feel we need to protect ourselves through safety in numbers*

> *Coming to know something through research, you feel a bit special because you are one of the few people in the world who has bothered to look at this. It isn't that other people couldn't do it if they tried. It is just that they have other things to work on.*

> *There is a lot of intellectual snobbery between different areas of mathematics ... I am not in an elite area, which doesn't mean that we don't think that what we are doing is interesting and important but analytic topology is looked down on enormously by everybody else in the field. I can see that in action in job appointments, research grants, the lot. I am not quite such an extreme case of that, but there is a bit of it.*

The men, on the other hand, focused on the 'other', either dismissively or appreciatively:

> *The relativity community as it usually defines itself seems to regard string theory as a bizarre construct which has nothing to do with what they would like to hear. This puzzles me. It is symptomatic of a certain kind of intellectual dishonesty, to be blunt. Relativists tend to regard themselves as intellectual descendants of Einstein.*

> *I have a lot of irritation with the domination of politicking in what is considered interesting and what is interesting research. It is difficult to know how you would eliminate it. People don't have time to read things properly so they pick up buzz words and what influential people are excited about. There is an extent to which everyone has to listen out for buzz words.*

> *I think it is perhaps the privilege of genius type people to talk about beauty and there are very elegant things that have been done by very profound people.*

From the anecdotes that were told to me, it is not unusual for women in mathematics to be the subject of sexist remarks:

> *In the first couple of weeks at university, one of my lecturers said, "Now you girls, I want you to be married to maths. I don't want you getting up on top of your wardrobes and saying go away, go away. This is to be a real marriage".*

> *Whilst the head of the group was very left wing and is married to a feminist woman, he is sexist and he couldn't quite believe that a woman could do the kind of stuff that they did. There was a definite prejudice. The second supervisor was very different. Again, pretty sexist but he was keen to have someone working with him so he did give me quite a lot of support.*

> *A younger colleague mentioned to me, when I got my PhD, when was I going to marry and he asked what else was there for me to do now I had my PhD?*

With reference to job applicants:

> *Looking at applications he had got in, he made a comment that if it was a choice between someone like me and a man with a family, he would give it to the man.*

This was the most overt behaviour recorded; it conveyed a message of limitation to the person addressed. All the essentialising components can be found in anecdotes such as those above; women are emotional, dependent, not quite up to it (whatever the 'it' might be), person-related, not highly focused, etc. At the same time, mathematics is 'objective', certain, systematic, etc. One of the things that such essentialising does is to remove the mathematics from the human sphere denying both the culture of mathematics *and* the mathematical culture. Essentialising fails to recognise that mathematicians, like everyone, live in a space bounded by the rules of the game of mathematics together with the rules of inter-personality.

When the female mathematicians expressed negative feelings, these were almost always about themselves: *I have learnt that I am more capable than I thought I was.* They put themselves down as *a plodder, over reaching myself* [in applying for a Chair], *believing that I was there because they'd made a mistake,* acknowledging that the *things I do aren't terribly important,* or *feeling a little bit embarrassed that I am not smart enough.* In a nutshell:

> *Mathematics makes one potentially very insecure. There is the panic when you know that you ought to know something and you cannot remember. There is the panic or uncomfortableness to do with not knowing, an underlying unease.*

Learners of mathematics also describe such panic (see, for example, Buxton, 1991). So, the message of alienation has certainly been heard by some. No male mathematicians made comments of this kind.

Supervision: When it came to describing their experience of the supervision of their PhDs, the negative images from women continued: *I didn't feel that I had made a great contribution [...] nobody seems to have a very high opinion of it.* Ten women and three men made comments which included reference to intimidation: *I never got beyond being intimidated by my supervisor presumably by practices such as having a really good idea in the middle of a supervision session, to jump in with that idea, leaving the student feeling extraneous; or being negative, I would say I tried this and he would say yes, I tried that five years ago [...] It was very depressing, competitive.* Arrogance, bullying, favouritism and the need for tenacity all featured. Three of them said:

> *Berkeley was a bit of a shock. It is enormous - it is like being at high school - there are hundreds of PhD students at Berkeley. On the one hand you get a diverse training, but nobody seems to care about you whatsoever.*

> *In those days they gave research students a pile of books and told them to go away and read them. The third year tutor was very patronising and she demoralised me. I felt uncomfortable and inhibited. This certainly had a detrimental effect on my work.*

> *There were 3 of us, one didn't come in very much and the other one was in algebra so we didn't have mutual interests nor support one another very much. It was quite lonely.*

Thirty-five of the mathematicians, 50%, made comments about PhD supervision not all of them negative, by any means. For example:

> *Doing the PhD was brilliant. I wouldn't say a word of criticism of my supervisor. He was exactly right for me and we got on very well. He had a lot of time. We had a fixed appointment every week but I would go in for as long as I wanted. He was very enthusiastic. He had lots of good ideas, was very encouraging.*

But she went on to say: *I know some people I shared an office with had very different experiences.* An example of such was a male mathematician who said:

> *My supervision was an awful experience because my supervisor was, in my judgment, an emotional bully. Eventually I submitted a theoretical physics/applied mathematics thesis plus statistics and operational research which was done unsupervised.*

This theme has been amplified in the work of Abbe Herzig. She commented:

> Overall, rather than defining instruction as an avenue to teach students to be mathematicians, these faculty described instruction as an avenue for students to prove themselves. This approach removes most of the responsibility for teaching and learning from the faculty, and instead places that responsibility almost entirely in the hands of the students, requiring students to develop a certain degree of competence *before* they can interact with faculty in meaningful ways. This program was structured to actively limit – rather than enhance – students' participation both in the program and in mathematical activity. (2002: 189-190 original emphasis)

In the interviews she conducted with 23 graduate students and 21 faculty members in one mathematics department of a large, public research university in the USA:

> students' interactions with faculty members during the first several years of graduate study mostly occurred while they listened to lectures in classes; even within the classes, there was often little interaction, either orally or in writing. The graduate students perceived what the faculty described: that the faculty were not interested in investing in the students until after they

> had proven themselves. The students described the limited and often negative relationships they had with faculty, and an almost complete absence of mentoring or advising. (Ibid: 201)

Like the mathematicians I interviewed, these students used negative words such as 'intimidation', 'pressure', 'poor teaching' and 'limited or negative relationships with faculty'.

I was interested to know how much the mathematicians had reflected on their own experiences of supervision when they themselves started to supervise. None of the 70 mathematicians had been supervised by a woman, but many of the 35 women were supervising and 3 of them, none of the males, made comments about connecting their own, with their students' experience. One said:

> *When I knew I would be supervising, I didn't spend a lot of time thinking about it. I hoped that it went with teaching and that it would come naturally to me but there were things that happened when I was a student which I made a mental note of at the time. I thought, well, when I'm a supervisor, I won't do that. I have begun to realize that things are different from the other side of the fence. I try to keep a balance between teaching, giving information, and leading my student the right place where he can find the information or do the learning himself.*

Another said *I don't think I reflected much on supervision at the time* but went on to describe how, now, she supervised much more tightly. In particular, she mentioned scheduling weekly meetings, trying to obtain outside deadlines for them to meet and getting them together, encouraging seminars and opportunities to *pick each other's brains*.

<u>Isolation</u>: Women often talk about the sense of isolation they experience, or observe others experiencing, within the world of mathematics: *Pure mathematics is quite isolating*. I noted that the women participants seemed to identify two kinds of isolation, personal and mathematical. Sometimes, they introduced isolation as a product of being one of a small number of women studying the discipline: *One thing I learnt from my own experience of undergraduate and postgraduate work was the isolating experience of being the only one,* and I am sure that that is an important issue for others. However, sometimes the isolation seemed to be induced more by the nature and practices of the discipline:

> *A statistical problem is never as isolated, it has all sorts of things impinging on it; it is always in a context. Whilst mathematics can be given a context it*

> *can be very isolated. Maybe women don't see themselves as isolated people.*

The participant quoted next used her personal learning experiences of isolation to influence her practices as a mathematician: *Almost all my research has been collaborative and I have got a huge amount out of the collaborative process. In my earlier studies I was denied this.*

Another mathematician, who had been discussing the differences between what I have called co-operation, inter-disciplinary teamwork, and collaboration with mathematical colleagues, confirmed the positive value of collaboration in part to reduce what can otherwise be the isolating effects of the discipline (see Chapter 7). One can imagine how potentially damaging, of the next mathematician, could have been the experience of someone else's isolation that she described:

> *In my second year [as a research student], I had a woman postgraduate tutor who was very bright. I admired her very much and I enjoyed the tutorials I had with her. At the end of the second year, I was completely shattered to find out that she was leaving the university because she found research hateful and lonely.*

Depressingly, the same words appear wherever the research is sited. And the results, too, are similar. In another study, a female student, who had just graduated from a prestigious mathematics department, commented in a research interview:

> *Can be very isolating. I enjoyed bits where I worked together with another person; this becomes harder and harder as you get more specialized. This is partly why I'm leaving – I love the maths but hate the isolating way of learning*

Community practices

The practices on which the women, and a few of the men, commented were located around hierarchies, competition and collaboration.

Hierarchies: Hierarchies are common in academe, often acted out as snobbery: " 'I have colleagues who are mathematical snobs of incredible dimension' " (mathematician quoted in Murray, 2001: 211). In the (British and Irish) world of mathematics (where my enquiries were situated), these hierarchies are role-legitimated (doctoral student, post-doc., lecturer, senior lecturer, reader, professor). A female mathematician explained:

> *The biggest hierarchy is the one that mathematicians are always putting themselves in. People rank each other.*

But hierarchies are also responsive to the assigning of value such as the difference between 'real' mathematicians and others. One woman said:

> *I consider myself a mathematician because I lecture and research in mathematics and I suppose that by definition makes me a mathematician. But I have met people whom I consider to be 'real' mathematicians and I am not one of those people.*

And Margaret Murray wrote of, and quoted, one of her interviewees, who were all female, using much the same way of distinguishing herself from others:

> she is careful to distinguish her work from 'real research' [...]which is the work of 'top mathematicians,' such as 'Gauss and Archimedes and Newton,' who reside 'somewhere near heaven.'. (2001: 214)

There is a very high value associated with being a Fellow of the Royal Society (very rarely, a female – a Fella perhaps?) or, even more, a Fields Medallist (the mathematical equivalent to a Nobel Prize in science). This persistent hierarchising and related evaluating leads the mathematicians to label themselves in relation to others. A senior, well-known mathematician said:

> *There is a level above the standard level at which we work. The top mathematicians can transcend these different fields and know enough. Often the dramatically new things are accepted at that level and then the smaller communities have to adapt themselves and, for example, learn the*

new language. Recently in complex dynamical systems very original ideas from probability theory have come in and have been accepted now by the world's top mathematicians and the complex dynamicists have to get on with it and accept it.

Hierarchies lead to promotional practices. And there appear to be rules to the game that work differently if you are female:

I applied for a chair at X. I was messed around. They short-listed three people. They had an external for each of the three areas of the three candidates. They were all male. The outcome was that it went to the guy in X area. I thought that one of the people was being aggressive with me. I came out feeling pretty bad but about two days later my Head of Department congratulated me on the tremendous performance and commiserated with me on not getting it. He said I had outshone the other candidates but the game had been stacked against me.

Amongst the many hierarchies in mathematics in the United Kingdom, Oxbridge has a uniquely powerful influence (see Chapter 3 for details). Of the surprisingly almost half of the participants who had been at Oxbridge as undergraduates, postgraduates or both, or were now working there having graduated elsewhere, many of the women, none of the men, were critical of their experiences:

Whilst I was at university I realised that I would really like to do research ... I didn't receive any positive signals in that direction and I was too shy to raise the issue while I was an undergraduate. Three years after graduation, I asked my moral tutor if she thought I could do research and received some encouragement and useful advice.

Earlier I reported that she continued:

The 3rd year [research] tutor was very patronising and she demoralised me. I felt uncomfortable and inhibited. This certainly had a detrimental effect on my work

Two other women said:

I went to Oxbridge to read mathematics. At [college], there were no tutors in mathematics so I had to go to [college] for tutorials. The tutor and I took one look at each other - I was very young and I found her horrendous. She was everything I never wanted to be. She was a complete battleaxe, totally unsympathetic.

> *Ultimately, my learning experience at Oxbridge (as an undergraduate) was all right but I could have drowned. It is a classic phenomenon [...] Retrospectively, Oxbridge taught me to be independent and individual but I wouldn't recommend the Oxbridge method.*

A male participant valued the very atmosphere that the females were regretting: *Oxbridge - the environment was a bit of a hothouse, which was good* and, in Chapter 3, we heard from the female about the negative experience of Oxbridge, expressed in terms of their not *do[ing] any nurturing. Your main source of knowledge was the hell of one-to-one tutorials. I used to feel ill every week before seeing my tutor.* She turned this into a positive experience by leaving, taking a break, transferring to another university and, once again, *really enjoying mathematics.*

Hierarchies not only permeate mathematics amongst university staff, but are also to be found in graduate student experience. Abbe Herzig (2002a) reported Etzkowitz et al. (1992) describing graduate research as:

> hierarchical and patriarchal. This is due, in part, to its origin in apprenticeship practices and a heritage of discipleship in which master scientists create successors in their own image as a form of asexual reproduction. (p.159)

It is easy to see how such mathematical cultural practices can result in the nurturing of some and the isolation of others and it is on such outcomes that Abbe Herzig's study provided evidence. Those who use apprenticeship as a model for learning (e.g. Brown, Collins & Duguid, 1989) do not discuss the gender aspects of this model.

But hierarchies are also to be found alive and well in schools. Elena Nardi and Susan Steward wrote about disaffection in the secondary mathematics classroom that they labelled T.I.R.E.D. (Tedium, Isolation, Rote learning (rule and cue following), Elitism and Depersonalization). Of elitism, they said:

> the hierarchy inherent in the above outlined elitist situation alters the nature of the classroom experience from one that focuses on catering for the individual learner's needs to one that focuses on establishing and assessing each learner's position in this hierarchy. The students express their alienation from this depersonalised, deterministic mathematical experience. (2003: 359)

The existence of hierarchies is one way of institutionalising competition but there are many other ways. Competition was something to which many of the mathematicians referred.

Competition: One female blamed the competition on the institutionalised teaching and assessing practices that are pervasive in schools and universities:

> *In some ways, I think I was a more intuitive mathematician before my more formal learning started, before university. All this being constantly measured and determined goals and examination, competition pushes you in a particular direction and makes you think you have done enough because you have achieved your goals. You could explore more. I am competitive and obsessive and I am not sure I should be encouraged in that.*

Another female said: *The competitiveness of mathematics is institutionalised from the beginning. So people are made to feel stupid if they don't achieve in mathematics* and male mathematicians were quoted in Chapter 7 describing competitive working environments. Apparently, engaging with competitive practices to promote your career in mathematics is a norm. A male mathematician gave some advice on how to do that:

> *People in mathematics who are more successful are people who check out markets and whilst it might be quite good fun being reasonably broad, you don't necessarily get a name for yourself doing that and I can increasingly see that it might make more sense to become more specialised. It is all about influencing people: whom you know, gaining respect for your work. It is important that you identify and, if possible, work with the people who are the movers and shakers.*

But being accepted, is more a social than an intellectual process, and is not quite so straightforward, as one female mathematician, a single parent, found:

> *I shared an office with my male colleagues but we didn't remain on terribly friendly terms. They went out drinking together. I didn't go to pubs. They formed a social group which was distinct from me. I thought of them as friends but not close friends and my friends were different.*

And, in any case, there was a level of *obsession* that did not appeal to these two women:

> *I suppose I feel that people whose life and soul is entirely within mathematics and they have a high profile on the public front I would consider to be mathematicians. People who are so engrossed in it that they*

can not see anything else in life. I don't think it is a particularly admirable state.

I found mathematics fascinating and I wanted to spend time doing and talking mathematics. But there was a choice between spending one's time with people who were fanatical about mathematics and only wanted to talk about it and people who were only really doing it because it was something to do and didn't really want to talk about it. I found very few people, no other women, who were interested in maths, wanted to spend time doing and talking about it but who also had another life. I didn't want to be totally obsessed with maths. There were too many other things that I enjoyed.

Going to conferences, and giving seminar papers, are two ways available to meet the *movers and shakers* and to promote your work but one female remarked:

I have been to too many maths. conferences where you realise that people in the audience haven't a clue what is going on and nobody thinks that is wrong. It has taken me a long time to realise that they aren't all brilliant.

So, the unspoken community rules are, again, in evidence:

The seminar is based upon the following structure - here is somebody's results, typically the analysis of some statistical data, here is why it is wrong, aren't they idiots; nasty - I can't cope with that.

Another female was reported in Chapter 7 talking about the seminar culture at conferences in which you make *a defence of your result against the audience which interrupts all the time, questions you aggressively, it is highly competitive. I hate that.* She continued: *There have always been a few who are very aggressive.*

Some women tried to break the rules: *I remember giving a seminar once and being told afterwards that I had made it look far too easy.* Perhaps, lying behind all seminars and articles should be the importance of convincing presentation. However, as became clear from the norm for presentation in mathematics seminars: *Some of the guys can't communicate but that doesn't matter to them.* Earlier, we heard from another woman who explained the contradictions:

I think that scholarly articles are often written to impress people how clever you are. I try not to do that. I don't like it. If you explain something well it can sound quite simple, but you don't want people to think it was easy because it wasn't. The same thing applies to seminars. So often they can put up hundreds of equations that you can't take in, you might have no idea what the person is talking about after the first five minutes.

Collaboration: Although almost all of the mathematicians (66 of the 70) spoke about collaborating in their research (and see Chapter 7), again collaboration is a social phenomenon and for some of the women mathematicians, it was not always smooth or satisfactory.

One of the things I found quite difficult when collaborating with my exact peers, was to do with the tensions, both social and intellectual.

I did have one four-way collaboration but the collaboration was extremely difficult and I don't want to continue with it and I think that was because we were treading on each other's toes. I think two of us were trying to play the same role and I felt challenged all the time. It was also too competitive and the wrong things were made too important. One of the collaborators was used to a very argumentative style and I hated it.

Not always sure how people manage to collaborate successfully. Sometimes I feel that I get put-down rather than helped.

Discussing a collaboration that functioned satisfactorily, another woman said:

what was important was that we weren't competing with each other. I wasn't a threat and we could collaborate easily.

Even when collaboration was working, there was a personal component for some:

I might try and formulate a question for a collaborator. That's marvellous when it works. I take a long time to feel safe enough to do that properly

Publishing

There is a macho view that the verbal is a corruption and a mortal danger; once you get into words, there is an associated lack of rigour!

Nonetheless, in the interviews, terms redolent of value judgments were freely used, as has been made clear in Chapter 8. The hierarchies on which I reported there, *interesting, important* or *significant* as labels for papers of publishing value and *trivial* for work that was not, were used by both genders but some women were less willing to decide what was 'significant'. One commented: *I probably leave it to the rest of the world to decide what is important or significant.* However, many of the female mathematicians, aware of just how value-laden such terms are, (and including one whom we heard, in Chapter 8, speaking about the hazards of the refereeing process), commented on their potential for misuse in the hands of some.

> *I am very much aware of a contrast between what I might find interesting and what the community of mathematicians to which I belong might, this year, find interesting...There is a problem that if you go and look for an area that no-one is working on, you might find such an area but it might be difficult to get something published.*

> *I think you can be novel and people will say so what. But I don't know how they know that. I think there is a body of mathematicians who make the rules up and I don't understand how they have made them up because they are part of the Old Boys' Club*

Indeed, the idiosyncrasies of referees were the focus of many comments and one female mathematician blamed *somebody who didn't know the subject* for rejecting her paper because: *Since then the same journal has published three papers with a subset of the results which were in my paper, by other authors.*

To be worthy of publication, a paper should be *novel* or *original*, show *potentiality* and, for some, be *timely*. Two male mathematicians commented on the race to publish:

> *There is much more machismo in mathematical physics; too many papers, published too quickly, everything has to be timely. Brilliant ideas which have to be rushed into print before someone else gets there first.*

> *Good mathematicians very often work in a very ambitious way [...] Anybody who survives will tend to have a stack of publications and, maybe for a short time, will tend to be known. That does not mean that you produce anything of lasting interest!*

But, only female mathematicians remarked on writing style. One said: *When I write anything, I think of the things I have suffered when people do not explain things well.* Three others explained negative practices:

> *I get annoyed with some of my collaborators and a lot of the papers I am sent which are definition, theorem, lemma, proof. That seems to me to be appallingly bad. It is the sort of thing that no-one is ever going to want to read. I think it is important to grab the reader from the opening sentence.*

Not "Let A be a class of algebras such that..." changed to "This paper opens a new chapter in duality theory". I write more colloquially perhaps; I think it is harder for people whose native language is not English to do that, but I do take slight liberties in order to grab the reader and say what the paper is actually saying.

Mathematics is full of hidden assumptions. You have to read a paper at least twice to get near the real agenda and then through a process of deduction to find out what on earth it is. Is it that the people who are doing it are blithely unconscious of the wider process or is it a deliberate process of mystification?

One of my collaborators likes to make things seem as sophisticated and complicated as possible and I think that is a problem for us. If I write the paper, our result, it looks like a simple result, and if he writes it, it looks like something else completely.

And yet, as Candia Morgan and I made clear (2000 and see Appendix C) although there is some published support addressing mathematicians' writing, they do not seem to be aware of, or recognise a need for, reflection on their, or their students', writing strategies nor of what might be accomplished by choosing to write in a different way. And it certainly does not seem to be something that the gatekeepers of publications, the journal editors, are addressing.

IN CONCLUSION

But it does not have to be as described:

When I did my honours year, we were all women, there were four of us and it was absolutely wonderful. It was friendly, it was supportive, it was intimate, it was co-operative. Everyone was so helpful and lovely - the thing I hated about doing mathematics was that the boys were so competitive and that competitive element fell away and we were just interested in helping one another to do mathematics. The tutors were all male and they thought it was great. They liked having us around, they liked having us in classes - we changed the atmosphere and made it possible for everyone to learn. My best friend went to Yale where half the students were female and they developed their own little eco-system. It was that year if anything that made me feel that I would like to be a mathematician.

It is possible that this woman's experience was particular although it is certainly at odds with the many other, conflicting, experiences described. However, it is also possible that some institutions manage to construct a cultural climate within which women feel much more comfortable. One example of such can be found in Rogers, 1990.

In the model put forward by Wenger (1998), there is an emphasis on duality. While participation and reification are, together, the basis of generating knowing through the negotiation of meaning, they are mutually dependent. You cannot have participation, for example in dialogue, without the reified objects of words. Further, "participation is essential to repairing the potential misalignments inherent in reification" (Ibid: 64) and vice versa. Under these circumstances, the derivation, existence and acquisition of such mathematical objects as formal proofs, computer programs or theorems are only possible through participatory practices in the community. If mutual engagement, joint enterprise and a shared repertoire are defined through the control of participation by a particular section of the community, then those remaining must conform if they want to be included (see Wenger, 1998: 73-85). Within the mathematical community of practice, the politics of participation have led to practices such as, for example, third person writing being reified into game rules. But

> to be effective, the politics of reification requires participation because reification does not itself ensure any effect. Reification has to be adopted by the community before it can shape practice in significant ways. (Ibid: 92)

It seems from the stories told above that such duality does not exist within the community of practice of mathematicians other than inside a very elite group. Furthermore, female mathematicians, sometimes knowingly, accept that this is the norm and adapt their behaviour accordingly. As Sara Ruddick argued women adopt "various strategies for coping with conflicts between reason and 'feminity'" (1996: 249). She mentioned:

I. "the universal standpoint of reason" (Ibid, 249) – 'objectivity' offers the best protection from "many sorts of misguided passion" (Ibid, 249). Some of the most staunch Platonists whom I interviewed would fall into this category.

II. Those who "temper a milder alienation from dominant ideals of rationality with a strong, positive desire to redescribe and endorse the cognitive capacities that have been labeled feminine" (Ibid, 250). Such women are difficult to find within the mathematical community.

III. "Many intellectual feminists [who] move ... between endorsing and rejecting both dominant and allegedly 'feminine' voices of reason." (Ibid, 250). Again, while I interviewed women who wished to change practices so that "they better represent and serve women" (Ibid, 250), they did not see such changes as being related to reason.

I suggest that one of the greatest successes of the mathematical endeavour has been to define and subscribe to:

> The ideals of rationality and objectivity that have guided and inspired theorists of knowledge throughout the history of western philosophy [and which] have been constructed through excluding the attributes and experiences commonly associated with femaleness and underclass status: emotion, connection, practicality, sensitivity and idiosyncracy. (Code, 1993: 21 quoted in Ibid: 248)

The engagement of females with such an epistemological stance is the more powerful because it invokes their 'informed' consent. As a result, the pervading epistemology of mathematical knowledge is claimed as universal. Under these circumstances, any links that it might be possible to make between knowing and knowledge are unavailable.

> No form of control over the future can be complete and secured. In order to sustain the social coherence of participation and reification within which it can be exercised, control must constantly be reproduced, reasserted, renegotiated in practice. (Wenger, op.cit.: 93)

It would appear that while these processes are outside the hands of many of those who wish to participate in the community of practice of mathematicians, such reproduction, reassertion and re-negotiation only take place within a selected group who are more interested in sustaining their current power than in raising questions about the health of the discipline and those who also wish to participate in it.

> From this position, what distinguishes irrationality from rationality, belief from science, or text from reality is not the cognitive level or type of the participating actants, but the associational enhanced power of some individuals and groups to establish resistant coalitions. Those who do are elevated to the labels of the rational, the scientific, and the discoverer of reality. Those who fail are the irrational, the believers, the followers. (Ward, 1996: 111)

And, one might point out, the relegated are likely to be female.

As Jerome Bruner indicated, it is impossible to separate knowing and knowledge, participating in coming to know as research or as learning.

> Our instruction in science from the start to the finish should be mindful of the lively processes of science making, rather than being an account only of 'finished science' as represented in the textbook, in the handbook, and in the standard and often deadly 'demonstration experiment'. (Bruner, 1996: 127)

However, recognising the relationship between knowing and knowledge as learning and as the outcome of research, and understanding that this is independent of who is the learner poses a serious challenge to those working in universities. It was very apparent to me that many of the mathematicians with whom I spoke made no link between their learning from research, and their students' learning nor between their students' learning and their own practices as teachers of mathematics. Indeed, they did not make the links that I have been making between the practices of the community and their experiences of their discipline. And yet their epistemological processes as researchers, contradictory to the knowledge epistemology driving the discipline, were as appropriate to the learning of their own students and of less sophisticated and younger students as they were for their own purposes. I believe, therefore, that our failure to teach mathematics satisfactorily or successfully to an heterogeneous community of students can be tied into the very situations that have been described. It is my belief that only when learning is understood as a result of research (see Brew, 1988) will it become possible to introduce and sustain effective practices amongst the community. And only at that time will the negative experiences described by many in this study be recognised as dysfunctional, for mathematicians *and* for the discipline.

SUMMARY

The mathematical culture is not one that celebrates diversity of members. On the contrary, the kinds of practices explored in earlier chapters of this book have been exposed in this chapter as operating to nurture some and discriminate against others. Where it is possible to influence or implement power, for example with respect to promotion, or to publishing, it seems that there are mathematicians willing to indulge in such practices. These practices, which often feature in some form of competition, are frequently institutionalised early in the school experience of mathematics but they are practices regretted by many, women and men alike. In the final chapter, I will discuss what are their implications for the learning and teaching of mathematics and how they might be changed.

That the women whose voices have been heard in this chapter remain engaged with their mathematics, and with their profession, is, I think, testimony to their strength, their commitment, their motivation, and, perhaps, to their patience! In Chapter 10, I appeal for a move away from a dysfunctional mathematical culture towards a culture that is equally productive for all who are part of it.

10

What does it mean to be a mathematical enquirer? – Learning as research

This book has reported upon a study that was designed to meet two main objectives. The first was to listen to what research mathematicians had to say about how they come to know mathematics in order to see the degree of match with an epistemological model that I had devised. It was my hope that a match with the epistemological model would provide grounds for justifying changes in the pervasive pedagogy to be found in classrooms. The second objective was to ask if the epistemologies, but also the practices, beliefs, values and attitudes of the female mathematicians were the same as the males and, if not, in what ways they differed. The recognition of similarities and differences between the experiences of the female and male mathematicians, I hoped, would reinforce moves towards a more socially just experience than I had reason to believe currently existed in research and teaching cultures. In Chapter 9, I made clear that the formation of female and male mathematicians meant that they did not differ in the range of positions that they occupied on the five categories of the model. However, there were substantial differences in how they experienced power in the practices within the mathematical community, and these were discussed in Chapter 9. Such differences have pedagogic, as well as experiential consequences, for those setting out to learn mathematics. In this chapter, I summarise the findings and explore their pedagogical implications. Aspects of the mathematicians' practices validate much current thinking in mathematics education about the teaching and learning of mathematics - a very welcome reinforcement that is underlined throughout the chapter.

In the process of doing the study, I uncovered surprises as well as confirming some expectations and these have been described in various chapters. However, what became clear to me was that the mathematicians spoke about their research enquiries in very different terms from how they described their teaching. In this chapter, I want to acknowledge these differences because I believe that the mathematicians' experiences, as learners, are relevant to less sophisticated learners in schools, and in universities. However, this belief, although supported by Brown, Collins and Duguid (1989), in which they drew, as exemplars, on Alan Schoenfeld's teaching of undergraduates and Maggie Lampert's teaching of fourth grade students (p. 37-39), is not uncontroversial. The mathematicians did not see themselves as learners, in the same sense that they saw their students as such; and supposed differences between creating new knowledge and learning existing knowledge have been discussed in the literature (see, for example, Ernest, 1998). However, elsewhere, I have pointed out that in order to learn, one is required to create and that apparent differences between researchers and learners

> are not a function of the site of the learning, research or school, nor the sophistication of the learner, but are a part of the climate for learning created by beliefs in 'objective' knowledge and the impact of those beliefs on classroom culture. (Burton, 2001b: 595)

Supporting the need for learner enquiry, a university physics lecturer said: "The process of learning is a process of invention and that ...means that if you're really going to learn something, then you have to invent it for yourself." (Marton, Runesson, Prosser & Trigwell, 1997, quoted in Bowden & Marton, 1998: 87). And one of the participants in my study reflected:

> *Whether what you are thinking about is new, research, known things or not, for you it is all new. When you understand a new proof, it becomes your own. Internally it is as though you did it. You feel you did it even if someone else did it. Internally it feels much the same because you have understood it.*

With my focus on learning, not on knowledge, I call on my own and others' research with learners of all ages and levels of sophistication to substantiate my arguments. To emphasise, then, differences in knowledge or in classroom or lecture teaching styles or organization are not my concern here. My focus is on learning about learning through reflection on the processes that support the learning done by researchers of mathematics, and comparison with the learning of students of any age, in any setting but particularly in schools or universities.

LEARNERS: ABSORPTION OR ENQUIRY?

The gap between practices of research enquiry and of traditional, transmissive classroom-based learning, I believe, to some degree helps to explain why learners find mathematics so difficult and unattractive. The Chicago Chronicle presented a conventional view of mathematics teaching and its impact:

> The traditional approach presents mathematics as a cut-and-dry proposition, that there is only one way to do everything - the way the teacher says. That approach has its drawbacks. In 1999, only 19 percent of eighth graders in Chicago Public Schools met or exceeded Illinois state standards in mathematics. Such performance in mathematics has dire implications for the preparation of the U.S. workforce. According to the American Management Association, 36 percent of job applicants tested by U.S. firms in 1998 lacked sufficient reading and math skills to do the jobs they sought. (2002)

Negative feelings and experiences in schools are indicated in the literature as widespread, and have been referenced throughout this book. Positive experiences recounted by students are recorded much less frequently. The absolute and fixed nature of the ways in which they meet the discipline is something that many students, such as those quoted in the research studies cited, find unacceptable and, indeed, that mathematicians, in their research practices, deny despite having had similar experiences as learners.

In Chapter 1, I referred to interviews I conducted in a study with students graduating from a highly respected department of a prestigious university. The students were asked what was the *best* thing about their undergraduate experience. Of the 31 students, 11 females and 20 males, only 6 females (roughly 50%) and 8 males (40%) made any positive comments about the course or their learning. Their other positive comments related to such issues as the prestige of the degree, the social opportunities, the 'cleverness' of their peers. I also asked them what they would have changed about their university experience, if they could. All, except one female who said she had no strong feelings, had suggestions to offer about aspects of courses, tutorials, lectures or supervisions and some of them reflected great anger. For example one respondent whose favourite subject, before university, had been mathematics and who qualified for entry to university with 4 'A's at Advanced Level said: *So it isn't as if I'm stupid or anything but I was certainly made to feel that way amongst fellow students and some of the teaching staff.* She summed up: *Not to be too cynical, I'd like the whole system to be changed but I think that that is practically impossible.* In Chapter 1, I also quoted a male student who reinforced what his female colleague was saying:

> *I think the way in which* [this university] *carefully selects the best candidates for each subject and then fails to fully exploit them is bordering*

> *on the criminal [...]all staff should be reminded that the aim of higher education is to make students able to use their subject.*

In the context of such remarks, it is salutary to observe that 54% of the females and 25% of the males in that study responded to the question of what they would be doing next, now they had graduated, by saying something along the lines of: *I intend to have nothing further to do with mathematics. I have found my experience here very demoralizing.* In the context of current pressures on students and teachers to perform, in mathematics, one wonders how the country can afford such departments of mathematics.

In case it should be thought that, with the new millennium, students' learning experiences have also changed, I can offer two recent (2003) pieces of evidence that the forces of resistance continue to flout the research evidence. The first was featured in a report in the New York Times Magazine, February 23, 2003:

> The release last month of a new math curriculum for New York City schools by Mayor Michael R. Bloomberg has elicited something just short of vituperation. Back-to-basics advocates denounce as "fuzzy math" its inclusion of so-called constructivist teaching techniques. Critics complain that those approaches encourage self-discovery and collaborative problem-solving at the expense of proved practices like memorization, repetition and mastery of algorithm.

The second is in the PhD dissertation of Heather Mendick. She noted the contents of a sign on a noticeboard between classrooms in a school where she had been conducting her research. The sign read:

> MATHS IS HARD!
>
> Independent research shows that Mathematics is the most challenging subject at A-level. Nationally, last year's AS results in maths were far worse than any other subject.
>
> If you don't really enjoy Maths and if you're not genuinely good at it, don't do it!
>
> Two years of struggling and constantly being "stuck" is not an experience we would wish on anyone.
>
> Success at A-level Mathematics usually depends on:
>
> **Positive attitudes**. Do you enjoy solving problems? Do you *like* Maths?
>
> **Persistence**. Do you give up easily and ask for help? Or do you prefer to get the answer for yourself?
>
> **Independence**. Do you need spoon-feeding every step of the way? Can you learn it by yourself?

The enquiry practices of the mathematicians give us guidance as to what might be done about this state of affairs. These enquiry practices are located within the epistemological model established by the research, but, in this book, I have discussed a culture associated with researching. This culture supports enquiry in ways that are very different from the cultural context of 'objective knowledge', the outcome of research where mathematics, separated from mathematicians, is presented as a supreme example of the triumph of reason over emotion. The inadequacy of such an approach to learning was explored by John Bowden and Ference Marton who pointed to "the paradox that we are trying to prepare students in institutionalized forms of learning for what is unknown (the future) by using what is known (our present knowledge)" (1998: 24) as well as to "the evil of the separation of knowledge from the acts and processes through which knowledge is born and is growing" (Ibid: 286) which I have called knowing, or coming to know. They said:

> The basic idea of learning is that the meaningful learning of concepts, ideas or principles has to be situated in real-life practices where these concepts, ideas and principles are functional and where they constitute discoursive [sic] resources for the learners. (Ibid: 58)

and that

> different fields of knowledge should include epistemological aspects of knowledge. 'Epistemological' is used here in a very wide sense, referring to questions about how knowledge is constituted, formed and how it is brought into being. (Ibid: 72)

The epistemological model confirmed by my study offers precisely this for mathematics, a structure that helps to constitute a culture in which mathematics 'is brought into being' within a discursive environment constituted by and with those within that community. Steve Koppes in the Chicago Chronicle (2002) described an example of learning in such a school classroom:

> The open approach to mathematics gives students some seemingly simple problems that illustrate key mathematical themes and asks them to solve the problems in as many different ways as possible. Students are surprised to learn that their classmates come up with very different ways of solving the same problems-so are the 25 teachers who received special training in the open approach to mathematics last summer.

Research-based examples to substantiate this have been quoted in this book. Further examples to be found in the mathematics education literature include, amongst many more, Boaler, 2000, 2002; Burton, 1999e; Davis, 1984; Fennema & Romberg, 1999; Jaworski, 1994; Morgan, 1998; Nunes & Bryant, 1996; Schoenfeld, 1985; Seeger, Voigt & Waschescio, 1998;

Skovsmose, 1994; Steffe & Gale, 1995; Willis, 1990. I am recognizing, in the ways in which the mathematicians come to know, an epistemological *and* pedagogical model which comes to terms with the "need at the very least to entertain the possibility that our most cherished beliefs might not only be wrong but be meaningless as well!" (Brown & Walter, 1990: 142).

IMPLICATIONS OF THE STUDY FOR LEARNING AND TEACHING

The model, which was the trigger for the study reported in this book, has five categories, person and cultural-social relatedness, aesthetics, intuition (and insight), different approaches, particularly to thinking, and connectivities. All five featured in the discourse of the mathematicians although not, of course, identically. Remembering that the categories of the model concern coming to know, not knowledge, below I address the implications for learning and teaching that they each carry, category by category. Except for connectivities, where all accorded importance to making and recognizing connections, whether within mathematics or between mathematics and other areas, heterogeneity featured around each of the other model categories. Mathematics itself, and those learning it, tend to be treated as homogeneous, even though many mathematicians write quite differently about the discipline. As Jo Boaler pointed out:

> Probably the main reason that teachers place students into ability groups in mathematics is so that they can reduce the spread of ability within the class, enabling them to teach mathematical methods and procedures to the entire group as a unit. (2002: 156)

And yet: "Various forms of evidence align to cast doubt on the effectiveness of ability grouping, certainly enough for us to think carefully about alternatives to this practice"(Ibid: 174). There is a widespread practice of attempting to make mathematics homogeneous; to abandon it for a recognition and incorporation of heterogeneity poses a challenge. The implications for teachers of acknowledging, respecting and working with heterogeneity are therefore also discussed.

The pedagogical implications of the categories of the model, taken together, contribute towards a classroom environment that I have described below in a section I have called Creating a Mathinking Atmosphere (after Mason, Burton & Stacey, 1982, Chapter 9). The learning implications of the practices, experiences and feelings, all of which were relevant to the mathematicians' understandings, are included in that section.

The chapter, and the book, concludes with a summary of the differences in mathematical culture experienced by the females and the males. I point out that such cultural differences are dysfunctional for the health of the discipline as well as those within it. I make an appeal to iron out both the current cultural inconsistencies as well as those between images of mathematics experienced by learners and those experienced by mathematical enquirers. I believe that this calls for a pedagogical approach to mathematics that treats learners as researchers.

Mathematics is related to the person in their socio-cultural setting

While I began this study with the conjecture that the mathematicians would be either Platonist or formalist in their epistemological views, I was surprised to discover a wide variation. When asked 'What is mathematics?' most gave predominantly Platonist responses but many also said something similar to this male: *I don't think philosophically about mathematics*, or this female: *I don't think about it. I am just curious and I do it.* Recent work in the sociology of science has pointed to theoretical mathematics as being "deeply affected by the history of modern societies" (see Lengwiler, 2000, reviewing Heintz, 2000).

Nonetheless, it is very encouraging that the uniformity that so often features in the public pronouncements of mathematicians, or of non-mathematicians, about mathematics, and that I reported in Chapter 3, has not been matched by the voices of my participants as they recounted their experiences as researchers. When it came to talking of learning, and particularly teaching, mathematics this aspect of the model also featured. One of the female participants said: *I very quickly came to accept* [through teaching] *that there was not one way of seeing mathematics.* One of the mathematicians provided a helpful outline of approaches that he arrived at through his experience of teaching. He said:

> *The one thing I have learned by teaching is that people learn in very different ways [...] Teaching is something that has taught me a lot about mathematics – about what maths is. I do quite a bit of service teaching and it is quite easy to forget about context, what maths is, what its history is, where it is coming from, where it sits in society and people's attitudes to it. I have found it very helpful teaching people with different views and different uses for it.*

Given this emphasis on heterogeneity, many students, as research quoted in these pages has shown, would welcome, and find very helpful to their learning, an approach of the kind that embeds the mathematics in its socio-cultural and historical contexts. But most of all, they would welcome being able to make use of the many different perspectives, knowledge-bases, interests and approaches present in their community of learners, when they are learning mathematics. Above, I have pointed to the contradiction between teaching a platonic form of mathematics, fixed and certain, and, in research, understanding *your* mathematics as an invention. But there are other contradictions associated with the mathematical culture, such as isolation and competition, that together help to sustain negative experiences.

Isolation and competition or collaboration within a community?

In Chapter 9, I reported on the isolating experience of undergraduate and postgraduate work as something that a number of female mathematicians mentioned and, in Chapter 1, I referred to one female graduating student who also hated the isolating way of learning that she had experienced. Speaking about when she was an undergraduate, one female mathematician said: *I had felt that I was being put down all the time and I am still quite angry about it. I could have done so much more with a bit of encouragement.* Although she is complaining of being diminished personally, such behaviours are not independent of a confluence of experiences from within the classroom culture. In this case, expectations for learning focussed on the individual may have led to competition, a product of which can be the kind of unacceptable inter-personal behaviours such as this mathematician is describing.

Above, and elsewhere in the book, I have outlined many of the negative experiences reported by students. These are reinforced in a report in The Guardian, 2003, of a study at Cambridge University done by Chris Mann in which she referred to "intellectual muscle flexing" in mathematics. She quoted one member of staff who described mathematics as "a kind of competition you train for" and that "women were unhappy with emphasis on speed in problem-solving. One student said: 'Women want to really understand mathematics rather than just crash through the examples' ". This repeats similar findings, also reported in this book, of research done in schools. I do not believe that either the country, or the discipline, can afford departments of mathematics that are having this kind of effect on students' learning. But, unfortunately, there are many such.

And yet, in Chapter 7, just over 94% of the mathematicians discussed collaborating or cooperating in their work and although such practices, occasionally, were not successful (see Chapter 9), their reasons for using them matched those given in the educational literature for encouraging students to work together. Two factors that were mentioned are relevant here. In a climate of inter-personal support, the mathematicians referred to a reduction in both isolation and competition.

Although many of the mathematicians made comments about their experiences of poor teaching, only one explained how her experience had stood her in good stead:

> *When I give a lecture to undergraduates, I tell them at each level how this relates to my own research. And you can see them prick their ears up. It all becomes very much more interesting and then they might go to the Library and find a book. And I think I got that out of lectures. I got the asides. I found out why people were doing that theory even if I hadn't understood the theory.*

If the heterogeneity and flexibility of the mathematicians could be part of the students' learning environment, if learners could be recognized as equally heterogeneous, encouraged and expected to act as researchers in their learning, and if the discursive, collaborative practices of the mathematicians were regarded as normal in the classroom and used to build a collaborative, inter-personally supportive community of learners, I believe that mathematics would become more accessible and attractive to students and that many of these negative behaviours would be undermined. Jean Lave explained the connections:

> This theoretical view emphasizes the relational interdependency of agent and world, activity, meaning, cognition, learning, and knowing. It emphasizes the inherently socially negotiated quality of meaning and the interested, concerned character of the thought and action of persons engaged in activity [...]this view also claims that learning, thinking, and knowing are relations among people engaged in activity *in, with, and arising from the socially and culturally structured world.* (Lave, 1991: 67, original emphasis)

One important part of the pedagogical shift I am advocating makes dialogue a feature integral to mathematics learning. Such dialogue involves talking with and about, not being talked 'at', talking to learn through the negotiation of meaning, not accepting the meaning of others. Used as a style, it emphasises for the learner that their identity in the classroom has shifted from dependency upon the teacher or text to *their* agency as a member of a supportive community. As Helle Alrø and Ole Skovsmose (2002) put it:

> Dialogic acts and dialogic interaction involve at least two persons in an equal relationship ... *a process involving acts of getting in contact, locating,*

> *identifying, advocating, thinking aloud, reformulating, challenging and evaluating*. (p.128, original italics)

In relation to the personal and socio-cultural perspective on mathematics, I have spoken with the voices of the mathematicians, and about them, in order to approach pedagogy of mathematics. I have addressed the contradictions between a classroom culture built around individualism, competition and isolation, and an 'objective' view of the discipline, as compared to a classroom culture utilizing the research practices of the mathematicians in the study involving collaboration, meaning negotiation, process as a necessary part of product that sees mathematics (at least during the enquiry process) as a human invention.

Teaching is not high on the agenda of mathematics tutors in universities but I have reported how, for one participant at least, important lessons were learned from a colleague when she worked in the USA:

> *I was teaching in parallel to an extremely experienced and popular lecturer and I learnt a lot from him and that was precisely the sort of thing, relating it to either his or their real lives*

While there is much more to teaching and learning than emphasizing relations between the mathematics and the familiar world of students, this is certainly one of the ways that more students can be encouraged to enjoy and acquire mathematics and to develop motivation and commitment to pursue the discipline.

But much more broadly, I am calling for a dismantling of a culture that is resisted by so many, and its replacement by the very culture in which so many of the mathematicians participate when they are researching. To change the mathematical culture requires more than an announcement that the class will, for example, do group work. There are many practices, behaviours and expectations that are covertly permitted, or even encouraged, albeit not necessarily deliberately, which help to build the culture. To achieve a change, a number of dysfunctional practices and stereotypes that have been refuted by this study must be dismantled. I am referring, for example, to individualism, competition, that mathematicians are born and that the mathematical culture is fixed. I continue to address this below.

Mathematics is an aesthetic discipline

The mathematicians were convincing in their passionate affirmation of the ways in which the beauty or elegance of mathematics brought them pleasure.

I am equally convinced that all learners, not only those operating at the fringes of current knowledge, are entitled to associate mathematics with positive experiences; and, indeed, such positive experiences evoke the kinds of feelings that the mathematicians expressed. I recall Kevin, aged 8, who had been having difficulties with a beam balance on both sides of which were the numbers 1-10. He had hung one weight on 10, on the left hand side, and was trying to hang 2 weights on numbers on the right hand side that, together, would balance it. With a whoop of delight, he suddenly threw himself at me from his seat on the other side of the classroom, shouting: *It's a 7 and a 3!* There was no doubt about his satisfaction, excitement, indeed euphoria, and I would challenge any mathematician to say that Kevin's discovery had less meaning for him, or was of less significance to him, than the research outcomes that the mathematicians reported to me. As I said earlier, this outcome emerged, for Kevin, because he was engaged in an activity that was meaningful and challenging to him where the epistemological framework was not centred on an objective-knowledge philosophy but in learner enquiry. His was the agency in his learning.

Learner Agency

What, pedagogically, can we learn from what the mathematicians said about aesthetics that can provide guidance for strategies to support a learning-as-research approach? A crucial difference between the mathematicians and most learners is that the mathematicians saw themselves as the agents of their own learning. Most learners are positioned as dependent on others, text authors, teachers, syllabus writers. This is despite, for example, one mathematician who said: *If you don't enjoy doing it, and value it yourself, there isn't a lot of point in always looking for confirmation from somebody else.* Agency invokes a sense of responsibility to yourself and those with whom you are working. Many students think that a good teacher is one who explains well. The mathematicians exposed a conflict between that view of teaching, often their own, and that of a good learner being one who asks questions, listens and responds to others' questions, helps to construct a convincing argument, makes connections, reflects and acknowledges, and where possible celebrates, the feelings thus evoked.

The culture of mathematics

The mathematicians associated positive feelings with aspects of the aesthetics of the culture of mathematics. They mentioned succinctness, compactness or conciseness, all of which help in generating and evaluating mathematical aesthetics but are also regarded as culturally functional. But the passage towards learning how to be, say, succinct is one that requires building and critiquing argument, developing experience of what, culturally, leads to one argument being labeled as 'nice' and another as 'nasty'. While examples can help, it is experience that is most powerful, and the association with positive feelings that can motivate is more likely to come from immediate experience. The same comments apply to other functional categories, such as simplicity, transparency and symmetry. Their power to convince is not immediately obvious to a naïve learner, nor indeed why these categories should be so valued, and not others. It takes experience, both of teachers and learners, to recognize and utilize them in an effective way and *enjoy the euphoria*. Such experience is not found in the majority of conventional mathematics classrooms. There is an overwhelming requirement in the education system that learners write to communicate, and they are judged on that writing. The lack of specific activities and experiences to support what the culture expects as well as the bases on which judgments are made should be addressed. Candia Morgan demanded:

> More explicit knowledge about the language that may be used to communicate about mathematical activity. Having knowledge of what forms of writing may be considered 'appropriate' and 'mathematical' – and what alternative forms of writing are available – is a necessary prerequisite for supporting student writing. (2001: 179)

As far as the motivational aspect of aesthetics, the mathematicians I interviewed expressed delight from the pleasure of touching perceived beauty and referred to how much such experiences motivated them. They were very sure that *beauty is in the eye of the beholder*, that is that people experience aesthetics in very personal ways but also that they found value in different aspects of mathematical practice and outcomes. Heterogeneity of this kind with respect to aesthetics is just as likely amongst learners.

Like everything else, appreciation of mathematical beauty is learned through experiencing the challenge, the struggle, the apparent arrival, the disillusion, the reconstruction, the conviction and sharing the associated feelings with those in the same community of learners. In other words, it requires learner engagement with activity, not passivity and dependency; there is an expectation that questions about that activity provoke and lead to learning. We are building, here, an image of a very different classroom.

In the report of her study of mathematics teaching through an investigative approach, Barbara Jaworski wrote:

> There has been evidence of high levels of mathematical thinking and challenge. Teachers seeking for intersubjectivity with students, trying to understand the students' thinking processes and offering appropriate challenges, was seen to be a significant feature of classroom interaction. The complexity of the teaching task lay in creating the sophisticated social environment through which meanings could develop. (1994: 213)

However, she warned of the non-congruity, in the case of one teacher, between:

> his planning and presentation of lessons [that] seemed to indicate an absolutist view involving the existence of invariant concepts which it was his task to deliver [and] personal concepts which individuals could be encouraged to develop, share and negotiate. (Ibid: 183/4)

She also gave a warning that must be taken very seriously when working for change in mathematics classrooms. 'An investigative approach' is not simple:

> It is likely that without a sound basic philosophy and sufficient clarity of vision (which includes a recognition of the influence of pervading cultures), changing to an investigative approach might have little chance of success. (Ibid: 185)

Part of the complexity of mathematics learning is located in the combination of the culture of mathematics *and* the mathematical culture. Aesthetics, for many of the mathematicians, were motivational, generative, evaluative and functional. As Barbara Jaworski pointed out, the challenge of incorporating into learners' experiences the expectations of cultural behaviours such as those described by these terms is certainly one of both philosophy and clarity of vision. It is epistemological and pedagogical.

Mathematics requires the use of intuition

In Chapter 6, I made reference to Rosemary Schmalz' definition:

> intuition is that faculty of the mind for which comprehension is spontaneous and immediate as opposed to rational and linear, and very often, though not always, sudden. (1988: 34)

as well as to her emphasis on what she called the disciplines of sustained attention, commitment, detachment and inner confidence which encouraged intuitions. What then is the goal of a mathematician who is also a mathematics teacher, as were most of the mathematicians whom I interviewed? Rosemary Schmalz spoke of three levels:

> At the lowest level the goal is to impart some applicable facts and to have these facts applied in simple situations...On a higher level, it is to teach deductive reasoning and proof construction. At an even higher level, it is to discover connections between facts that are later verified in an accurate proof, or to solve a problem by creative use of some applicable facts. Thus, the goal is not simply to pass on some tools useful in applications; it is not simply to pass on a set of useful problem-solving techniques. It is also to create situations where students will discover the power of their intuition. Perhaps the foremost goal of a mathematics educator then is to construct situations where students may experience small intuitive breakthroughs. (Ibid: 42)

For me, there are two important issues raised here with respect to the learning of mathematics. The first is about attention, commitment, detachment and inner confidence. Do mathematicians and mathematics teachers set about ensuring that students engage with these "disciplines"? In my experience, this happens rarely and not in a sustained way although, above, one of the mathematicians described how she did deliberately set out to do so. But unfortunately the necessary passion and energy are not found in most mathematics classrooms and students, on the contrary, are given to understand that their feelings are not integral to their learning. But this is the antithesis of the approach to research of the mathematicians to whom I spoke. Experiencing "small intuitive breakthroughs" is dependent, then, on students being put in situations where they experience, and reflect upon, the power of such breakthroughs in order to learn how to access that power. This cannot happen in a classroom that is 'delivering' facts or tools where the purpose is defined by being a required part of a syllabus. The attention and commitment of the students to the activity on which they are engaged is a necessary part of ensuring that they experiment with recognizing and using their intuitions. This, in turn, contributes both to their inner confidence and to learning the importance of being able to detach from and look at a situation they are trying to resolve and make judgments about the useful strategies they might adopt. All of this requires that the students are actively engaged with mathematical problem solving and not with repetitious or rote-learning practices.

A second issue, for me, relates to Rosemary Schmalz' goals with which, I feel confident, she would have the agreement of the participating mathematicians. If the overriding goal is to facilitate the experience of "small intuitive breakthroughs", this is unlikely to happen in a classroom which (a) does not overtly have that as a goal for teacher *and* students and where (b) the "lowest level" goal, imparting and applying facts, dominates the pedagogy.

Some of those mathematicians who talked about their students with respect to intuition, did so in somewhat dismissive terms:

> *One of the things I find about students, undergraduates in particular, is that they seem to have very little intuition. They are dependent upon being spoon-fed. The ability to look at a problem from different angles is crucial. They tend to look from one angle and you cannot see a way through. If you look at it from different angles, mysteriously perhaps, something dawns on you and you find a way through.*

There is no sign here of any responsibility for a sense of dependency in students, or for a recognition of the need to model the kind of behaviour that this mathematician himself finds so constructive. This is consistent with the report that 68% of the students in a 'most-able' class in a study, possibly future mathematicians, prioritized memory over thought (Boaler et al., 2000: 637) confirming Jo Boaler's (1997a) finding from a previous study that 64% of the top achieving pupils believed that remembering was more important than thinking. Kathryn Crawford and her colleagues (1994) found that 82% of university mathematics undergraduates whom they studied used a reproductive strategy to learn. Reproductive strategies are not only memory-dependent but they are also a product of a dependency culture in the classroom in which knowledge is located in the text or the teacher, and is outside the agency of the learner.

However, one of the female mathematicians in the study took a different perspective:

> *PhD students don't have the depth of experience to have such realizations. I think you have to help them by posing the questions and leading them. You model the process. I think maybe you explain it as you are going along; you try and lead them along the way so that they experience looking for the answers.*

Vera John-Steiner appealed for:

> instruction in diverse problem-solving strategies just as there is instruction in varied laboratory techniques during the apprenticeship of young scientists; however, this is seldom the case. Instead young men and women are exposed primarily to what Medawar has called 'the art of the soluble'. (1997: 182-183)

She reported a question that she had posed to a physicist, John Howarth, of "whether it was possible to help students to rely upon intuition as part of their preparation for becoming physicists". He replied:

> Intuitive solution of problems is important. Essentially it is finding the answer to a problem before you have solved it. Students are tempted to believe that physical intuition is something that you either have or don't have. We certainly all have different talents, but the process can certainly be encouraged – that's one of the things that teaching is about. Teachers can encourage the talent by example and by describing their own approach to problem solving. They can also take the time to explore the student's process with him or her. (Ibid: 183-184)

And Smith and Hungwe pointed out: "If guessing and the resulting cycle of inquiry does not become visible to students, they are left with only public mathematics – the carefully crafted propositions and polished arguments they see in their texts" (1998: 46). Jon McKernan asked:

> What's so awful about using intuition or using inductive arguments? [...] without them we would have virtually no mathematics *at all*; for, until the last few centuries, mathematics was advanced almost solely by intuition, inductive observation, and arguments designed to compel belief, not by laboured proofs, and certainly not through proofs of the ghastliness required by today's academic journals. (1996: 16, original emphasis)

And yet, except for Efraim Fischbein's (1987) book, the literature of mathematics education does not address intuition as a focus of concern in the classroom and neither did the mathematicians appear to consider it their responsibility to nurture intuition in their students. But, if the mathematicians are correct and their intuitions are a product of a combination of knowledge and experience, *both* are educable. Furthermore, from the mathematicians, and the authors quoted, we have clear indications as to what is important. As one might expect, pattern-searching featured prominently but so also did using metaphors, and drawing analogies. Making connections was also emphasized. These are aspects of how one acts, *qua* mathematician, not about the particular aspect of mathematics in which one is engaged, or the particular mathematical relationships one is trying to establish.

In Chapter 5, I mentioned the pedagogic contradiction that asserts the importance of intuitions to mathematics but does little to nurture them in students. Dealing with that contradiction is a necessary part of developing mathematical 'sense-making'. Not only does that require a teacher calling for intuitive responses to an activity, and encouraging them when they are made, but also, when an intuition proves to have been misguided, helping students to look for why. It demands the recognition of connectivities, overtly, in classrooms. Actively asking the students of what does something remind them, in what ways are things similar and different, helps to expand their potential for making intuitive leaps as well as building the complex map of mathematics to which the mathematicians referred so positively.

There are three different thinking styles used in mathematics

One of the biggest surprises of the study was to discover that the two mathematical thinking styles featured in the literature, visual and analytic, were insufficient to describe the ways in which mathematicians thought. We need to include a third, that I have called classificatory thinking. Sometime after the study, on hearing me describe these three styles, a colleague in Germany translated for me a letter from Felix Klein in which he distinguished mathematicians as:

> 1. The philosopher who constructs on the basis of concepts.
> 2. The analyst who essentially 'operates' with a formula.
> 3. The geometer whose starting point is a visual one.
>
> (See Ferri & Kaiser, 2003: 213-214.)

While my study would not necessarily support his description of the different mathematicians, philosopher, analyst and geometer (as, for example, I found many examples of those who thought visually but not strictly-speaking about geometrical material), I was delighted to be able to provide the empirical evidence to support agreement between my description of thinking styles and that of such an august mathematician! Klein's classification arose from observations resulting from co-operating with other mathematicians. He allocated himself and four colleagues to one of his three types; one (Kronecker), he labelled between types 1 and 2. (See Ibid: 214.) In my study, 25 of the 70 mathematicians appeared to use one thinking style only and only 3 appeared to use all 3 (see Chapter 4). Amongst those whose thinking style was visual, there appeared to be little acknowledgement of other ways of thinking about mathematics even when they used one of the other two styles. Those whose thinking style was *not* visual seemed to have a feeling that this made them not quite top class.

A number of observations with educational implications result from this. First, the work by John Cowan (1975), cited in Chapter 4, suggested that thinking and learning styles in engineering education are inter-connected. More importantly, his work indicated that an engineering lecturer with one dominant thinking style is likely only to be communicating fluently with those students who share that style. My work, in mathematics education, reinforces the importance, to learning, of both teachers and students becoming sensitive to differences in mathematical thinking style and their potential impact on the making of mathematical meaning. I am not aware that such differences feature in courses of teacher education or in the thinking of teachers at any level of the educational system. In Chapter 4, I mentioned the social stereotype, also held by mathematicians, that visualization is important to mathematics but I have not met teachers who reflect on whether or not *they* are visualisers and what impact this might have on how they present and talk about mathematics, the images that they use to support their thinking and, indeed, on whether their students need and do likewise. Certainly, as a student of mathematics, my thinking style was never something on which I reflected nor did any of my teachers cause me to think about how my thinking style was influencing my mathematical development, to consider alternatives, nor help me to develop strategies that might be an influence.

The second point is that becoming aware of differences and reflective about one's own strategies is very likely to influence how, as a teacher, one deals with students. I reported in Chapter 4 that John Cowan confirmed that the engineers, who participated in his study and reflected on the findings, did change. It seems likely that an aware teacher would attempt to use different perspectives when presenting material to students on the assumption that a student's failure to understand is possibly caused by something like lack of similarity in thinking style rather than inability to learn. At present, I doubt that it is understood in these terms. From the students' perspective, becoming cognizant of these differences must surely expand their mathematical views and give them more strategies for working mathematically.

Third, teaching materials are not constructed to exploit differences between thinking styles and offer alternative views, pathways, for reaching a particular mathematical goal. Curriculum developments that alerted teachers to differences in thinking style and provided opportunities for learners to exploit their dominant style and learn from its distinctiveness, as well as be aware of its disadvantages, would be more mathematically insightful than repetitive presentation of the same mathematics. This would seem to me to be far more important in an agenda for teacher education than the detailed specifics of the mathematical syllabus objects.

Finally, I want to draw attention to the fact that the mathematicians usually had many problems on which they were working simultaneously, that a lot of that work was done informally, away from their desks, that they regarded errors as normal, that they frequently became stuck and, when they did, moved from what might be causing this state, to something different in order to unblock their thinking. The strategy of a student working on more than one problem at a time, almost unheard of in mathematics classrooms, and of having time and space to retreat, reflect, research, is not only appropriate to the unsolved problems of research mathematicians. Students undertaking a mathematical challenge also need to have room to manoeuvre, to work together, to consult people or books, to think. Most of all, instead of being overwhelmed by frustration when stuck, students would benefit from knowing that mathematicians find it "an honourable and positive state, from which much can be learned" (Mason et al., 1982: 49) and more than that: *mistakes and errors are part of mathematics. You cannot live without errors in mathematics.* And yet, as teachers, the mathematicians did not necessarily have, or want to model, strategies to help their students. One mathematician reported:

> *My supervisor was the sort of person who expected his graduate students to stand on their own feet. If you went to him and said you were stuck, you got silence. If anyone was going to un-stick you, you were going to do it yourself.*

While I would certainly not advocate that the agency of a graduate, or indeed any student, should be undermined, the knowledge and experience of a supervisor or a teacher puts them in a special position to be able to make their students aware of strategies that are productive for dealing with, amongst others, a stuck state. I believe that it is unprofessional to abrogate that responsibility. We need to construct a classroom atmosphere that incorporates stuck, as a natural state, and the legitimacy of walking away from it as well as recognizing, using, comparing and contrasting the influences and outcomes, including errors, of the three different thinking styles.

Mathematics is a network of connections and is, itself, connected to other areas

Only when it came to speaking of connectivities within mathematics and from mathematics to other areas was there almost universal agreement as to their importance, whether or not the mathematicians could make connections to their current work:

> *Sometimes one comes across the missing link. A small idea closes up a circuit that has been around not closed up for a long time.*

> *Your work is not geared immediately to applications although it would be excellent if one could establish the bridgehead which allowed that.*

> *Some ideas that were understood by one group of people were exactly what another group of people needed to make something work.*
>
> *You feel that what you have been doing is part of something bigger.*

Whatever the specialty in which the mathematician was working, they were convinced that mathematics constituted a network of knowledge, not separate pieces. Even if the connections were not immediately visible at present, they were sure that connections would be found at some stage. As applied mathematicians, their priority was not necessarily finding such connections but nonetheless they acknowledged their importance.

All the more regrettable, therefore, that mathematics is presented to students in fragments. Whether these fragments are constituted in textbooks with a chapter on one area, followed by another chapter on a different area, or in courses where the lecturer concentrates on the particularities of her/his area frequently to the exclusion of references to connectivities, the message to the students is one of disconnection, not connection. This makes unavailable to students that powerful mathematical tool of searching for analogy, of asking: Have I seen something like this before? So students are denied the connected view, despite agreement amongst mathematicians that mathematics is a network of knowledge. I find this incomprehensible and malicious since learning through connections, as the students above said, invokes understanding and is *not* memory dependent in the way that learning disconnected fragments has to be.

Where do 'the basics' fit into this? The working practices described are built around the solving of problems, not the learning of disconnected objects of knowledge. The mathematicians were clear that when they began a problem in an area that was new for them and they had, as it were, to learn 'the basics', they did this by engaging with the problem and searching for ways of understanding and deconstructing it: *To solve a problem, you have to go and find out about some maths that you didn't know.* There are no 'basics' here. Part of the drive is to identify what is, and is not, useful, in the context of the problem being addressed. As was demonstrated in the study conducted with 16-year-old students, to which I referred in an earlier chapter, those students asked for the same:

> *More understanding than memory because, if you understand how to do something, then you'll be able to go back and do it again.*
>
> *If I don't understand, then I can't - although I might be able to do it. Then if I don't understand it I get confused as well so I have to have both, be able to do it AND understand it.* (Students quoted in Burton, 2004, forthcoming).

The teacher is posed with two different ways of dealing with connectivities. First, through the construction of courses, and supporting text, teachers should be sensitive to the building of connections and do so with determination. Second, when students are working at solving problems, they can be regularly reminded to ask themselves, and one another, about similarities and differences, inside and across areas, and the possibility of an analogy being productive.

Mathematics and those learning mathematics are heterogeneous

The assumption of the *homogeneity of knowledge* states that it is possible to integrate all sorts of knowledge into one unified system (Skovsmose, 1994: 196, original emphasis). This

> brings into being an assumption of the existence of an *authorized body of knowledge* (a body of 'truths') [...] The possibility of different competing bodies of knowledge is ruled out, and 'knowledge' comes under control. (Ibid: 198, original emphasis)

The consequence of this, as I have pointed out earlier, is that the teacher is responsible for passing authorized knowledge to the students; hence transmission teaching has become the norm. However, the epistemology of the mathematicians, as evidenced in how they research according to their own statements, supports a view of knowing which makes

> A critical investigation of a knowledge claim [...] a permanent necessity. There is no knowledge without preconception and prejudice. We have no foundation, no homogeneity and no authorized body of knowledge. We have to introduce a perpetual critique. (Ibid: 200)

Once the process of critical enquiry is recognized as the best way of exploring mathematical knowing, to make the necessary critiques requires an incorporation of the mathematician, researcher or student, the person, into the process. Thus, there is a need to involve "the learner as an agent in the learning process. This establishes both 'intention' and 'reflection' as educational concepts" (Ibid: 201). But the finding of heterogeneity in approaches, beliefs, practices of the mathematicians also offers a challenge to recognize the richness of variation amongst all learners. The variation helps to provide evidence in the form of counter-examples (as shown in Chapter 3) to counteract the myth that you are born to excel at mathematics. But nothing is richer as a learning medium, than a context of variation. Ference Marton and Shirley Booth

(1997) explained "the variation in how most [learners] make sense" (p.163). They pointed out that individual learning "can be understood as an appropriation of successively larger and larger regions of the object of learning as described on the collective level" (Ibid: 163).

Heterogeneity, then, is integral to the mathematics as well as to those learning it, as we have seen with the practices of the mathematicians. This requires very different intentions and behaviours of learners and of teachers. Transmission and reproduction are recognized as dysfunctional to learning. "The step from a monological to a dialogical epistemology is a way of making impossible the notion of knowledge as authority and of education as 'delivery' " (Ibid: 206). Below, I explore the implications of creating what I have called a mathinking atmosphere which, I hope, would help to ensure that mathematics is an acknowledged part of everybody's cultural experience and not, simply, a legal entitlement or economic necessity.

CREATING A MATHINKING ATMOSPHERE

In this book, I have provided the evidence for a substantial shift in mathematics education (that is, the teaching, learning and assessing of mathematics) away from a content, objective-knowledge perspective, to an appreciation that mathematics is learned via a process, both epistemological and pedagogical, that is often lengthy and rarely smooth. This process, amply described by the mathematicians I interviewed, depends upon personal and socio-cultural active engagement, the resources of a community, the recognition of the interplay of emotion and cognition, a valuing of heterogeneity and an appreciation, through reflection, of the complex ways in which mathematics connects, internally and externally. As Tony Ralston pointed out to me, understanding what he called the "mathematical enterprise" is more important than knowing quantities of facts or skills, that is, it is about how we engage students in the activity of mathematics, not about how much they learn (private communication). But it also depends upon a dismantling of those features of the mathematical culture that currently operate to facilitate and sustain some members of the community, and disadvantage others. To learn mathematics, I am asserting, requires an atmosphere which takes account of these features.

One way to understand the kind of shift that I am recommending is to take a narrative perspective on the learning of mathematics. We come to understand through narrative, telling stories, listening to and acting on stories to make our own meanings. The intention of such utterances or texts "is to initiate and guide a search for meanings among a spectrum of possible meanings" (Bruner, 1986: 25). In this book, we have been presented with the narratives of mathematicians about how they come to know. Each of these is, of course, important to the person telling the story but from our perspective as listeners, we make meanings by looking for similarities and differences between the stories in order to construct a meaningful story of our own. It is out of the confluence of many of these stories, reinforced by their similarities, that I have offered a narrative that distinguishes between the messy process of coming to know, and the tidy public presentation of mathematical knowledge, between an epistemology of knowing and one of knowledge. I have offered the epistemology of knowing, what Ole Skovsmose called "reflective knowing" (1994, Chapter 6), as a reasonable guide to learning and emphasized the links between coming to know through the meanings found in narrative and a pedagogy of enquiry, of the kind experienced by the mathematicians.

Elsewhere (Burton, 1999d), I drew on Jerome Bruner's (1986) distinction between imaginative and paradigmatic narratives, in order to make clearer the mathematics education priority of facilitating knowing, rather than acquiring knowledge, in order to make mathematical meaning. I advocated that mathematical meaning is best made when the agency for learning rests with learners and when they acquire and use the practices of authorship. In recognizing this, one of the mathematicians said: *As a teacher what I have to do is show, for a learner something has to happen in them.* Processes such as the exploration of conjectures, by creating, testing, falsifying and validating, the setting of boundary conditions that influence whether conjectures are, or are not, valid, the constructing and challenging of argument and the deliberate use of reflecting, all feed agency and authoring. The responsibility is the learner's, and the narratives that are told have the purpose of offering meaning to the individual and the community that, then, becomes a focus of re-interpreting, challenging and convincing. "We always have to make a knowledge claim an object for further investigation. We have to negotiate what is conceived as the content of the knowledge claim and what is the basis for 'being sure' " (Skovsmose: 1994: 201). But process and content are not in opposition, or even separable, so we must not be led down the cul-de-sac of separating them:

> Although Schoenfeld may appear to be teaching strategy rather than subject matter, he was, more fundamentally, building with his class a mathematical belief system around his own and the class's intuitive responses to the problem [...] working in the culture of mathematics, not in the culture of schooling, he did not have the students stop at what, in culture of school practice, would mark the end: an answer. (Brown, Collins & Duguid, 1989: 38)

I am asserting that the process of authoring meaningful mathematical narratives is a necessary part of learning at every level and cannot be replaced by the apparently simple transmission of facts or skills. I believe that this is what "working in the culture of mathematics" is about. But, in Chapter 8, I discussed the lack of agreement about authoring practices amongst mathematicians and, more worryingly, the failure to address the cultural style practices and to tease out their effects. Since authoring, both informally and formally, is an important part of communicating with, and convincing, the community of the worthwhileness of some mathematical work, it is about time that such issues were addressed with learners, as well as by researchers themselves. (For a very helpful approach to this in secondary school classrooms, see Morgan, 1998 and with respect to the mathematicians' writing see Burton & Morgan, 2000, Appendix C.)

In Chapter 6, I quoted Brian Rotman (2000) who described the need, that is the motivation, of the mathematician to follow and be convinced by an argument, to "seek *the idea behind the proof*" (p.18, original emphasis). He went on to point out that:

> It is perfectly possible to follow a proof, in the more restricted, purely formal sense of giving assent to each logical step, without such an idea...attempts to read proofs in the absence of their underlying narratives are unlikely to result in the experience of felt necessity, persuasion, and conviction that proofs are intended to produce, and without which they fail to *be* proofs. (Ibid, p.18, original emphasis)

Brian Rotman called "the linguistic resources mathematics makes available" the mathematical *Code* and meant "by this the discursive sum of all legitimately defined signs and rigorously formulated sign practices that are permitted to figure in mathematical texts" (Ibid, p.18-19, original emphasis). He warned that:

> mathematicians ...when moved to comment on this aspect of their discourse might recognize the importance of such narratives to the process of persuasion and understanding, but they are inclined to dismiss them, along with any other 'motivational' or 'purely psychological' or merely 'aesthetic' considerations, as ultimately irrelevant and epiphenomenal to the real business of doing mathematics. (Ibid, p.18)

However, he called the *meta-Code* "the penumbra of informal, unrigorous locutions within natural language involved in talking about, referring to, and discussing the Code that mathematicians sanction" (Ibid, p.19). It was in the meta-Code that he located "the idea behind the proof". I believe that it is in this gap between Code and meta-Code, or, in my (educational) terms, between the transmissive texts (the coded, paradigmatic, narratives) and the very different practices through which the content of these texts is explored and understood (the informal, imaginative, narratives) that explanation can be found for the use, across the world, of the unconvincing process described above by Brian Rotman in classrooms. Most students do not "feel necessity, persuasion and conviction".

A mathinking atmosphere is one that supports enquiry. And enquiry depends upon asking questions, constructing argument and critically reflecting on outcomes: "mathematical thinking [...] offers a way of structuring, a direction of approach, a reflective power as well as creative and aesthetic potential [...] resolving brings a sense of pleasure and confidence" (Mason et al., 1982: 155). That is, the existence of a community of learners sustains enquiry. Unfortunately, too often mathematical classrooms are fact-, or skill-dependent, operating within an individualized, rather than a collaborative, paradigm. Robert Thomas has pointed out that "Mathematical facts, without some understanding of why they are the way they are, are almost impossible to learn and too boring to keep awake for" (2002b: 45). In these pages we have heard the voices of students saying exactly this but we have also been presented with evidence that such voices are not heard.

The mathematical enquirers who contributed to this study have drawn attention to the centrality of problem solving to learning. In turn, this approach emphasizes questioning, exploring, conjecturing, testing, validating, falsifying, and convincing, all in a community of learners as a social, not an individual, concern. *I am sure that students can benefit from working on problems together. It is how I learnt - it can be very helpful talking to your peers.* But we are not simply talking about putting students into groups to work together. There is a need to build a participatory learning community (Wenger, 1998), to tease out, support, and reflect upon the complex ways in which knowing is built among and between members of that community. "Within a culture, ideas are exchanged and modified and belief systems developed and appropriated through conversation and narratives; so these must be promoted, not inhibited" (Brown, Collins & Duguid, 1989: 40)

Learning in a social context depends upon the mathematical culture that defines, influences and is influenced by the members of the community. Unfortunately, the mathematicians, particularly the females, talked in terms of arrogance, smugness, intimidation, pressure, isolation, competition, the role of hierarchies - the exercise and abuse of power. Negative experiences are often located in reified, competitive statements about mathematical objects (for example in the seminar behaviour reported in Chapter 9), or put-downs on a personal level (also reported in Chapter 9). I am suggesting that an absolutist view of knowledge reinforces, and is reinforced by, the discourse of power; hierarchies are endemic to such an approach. It operates so pervasively that it is no wonder that the disease is caught by students and transmitted when they, themselves, become teachers. Many do not like, or approve, of it. A shift to a learning perspective that is enquiry-based would, I believe, not only undermine this destructive culture but, because it would require an epistemological re-positioning that incorporates inter-personal respect and collaboration, would be welcomed by students. And this would be consistent with the mathematicians' current collaborative practices in their research.

As has been clear throughout these pages, the participants in this study freely acknowledged their own short-comings and needs, as well as those of others. *In academic life, there is an underlying assumption, which is wrong, that teaching is something you can do.* And yet, many students complain about the teaching and the lecturing to which they are subject: *The permanent faculty are often slated by students for their teaching.* Hyman Bass wrote about this situation and the obligations that it imposes:

> The time has come for mathematical scientists to reconsider their role as educators. (Bass, 1997: 19)

> The disposition of many mathematicians toward the problems of education well reflects their professional culture, which implicitly demeans the importance and substance of pedagogy. Mathematical scientists typically address educational issues exclusively in terms of subject matter content and technical skills [...] Pedagogy and content are inextricably interwoven in effective teaching. (Ibid: 20)

> Mathematics education, unlike mathematics itself, is not an exact science; it is much more empirical and inherently multidisciplinary. Its aims are not intellectual closure but helping other human beings, with all of the uncertainty and tentativeness that that entails. It is a social science, with its own standards for evidence, methods of argumentation and theory building, professional discourse, etc. It has an established research base, from which a great deal has been learned in the past few decades that has an important bearing on the educational performance for which academic mathematicians are responsible. (Ibid: 21)

Mathematicians are part way there in that they already re-position themselves with respect to their own research enquiries. There remains a need to convince them of the applicability of this stance to their own teaching, and to the potential for learning of their students. Once they are convinced and the practices changed, there is a route, through their students, into schools.

CONCLUSIONS

An epistemology is a theory about knowing, and the model that I developed robustly describes knowing, in the experiences of these seventy mathematicians. Knowing is understood as socio-culturally based, as being aesthetic and intuitive, heterogeneous and holistically inter-connected. Knowing and feelings play on, and support, each other. Within this view of mathematics, the discipline does not 'belong' to those who work within it, but it is part of the cultural heritage of all. Whatever the level at which students enter the discipline, they are entitled to walk away from their encounter empowered by mathematics. Encountered in these pages in the narratives of the mathematicians are images of mathematics that carry excitement, creativity, beauty, and many other positive features. The images of mathematics reported by learners do not. I have suggested that the gap between the mathematicians' views of mathematical knowing and that encountered by learners is monstrous. It could be said to be an indicator as to why so much teaching of mathematics fails in that, currently, it comes from philosophical and epistemological perspectives that are unsupportable in research terms and disconnected from the enquiry experiences of, amongst others, research mathematicians.

I do not want to accord definitive power to the mathematicians. What I am suggesting is that enquiring mathematically is the normal way to establish knowing. Understood as learning, these experiences are the same whether one is a more experienced or more naive learner. Wanting to know, i.e. understand, and make use of mathematics, as opposed to pass a test or gain a certificate, is a necessary part of engaging with the discipline, enhancing one's familiarity with its techniques and knowns, and searching for connectivities with other knowns of one's own, one's peers, within or without the discipline. But it can also be seen that the unacceptable social practices within the discipline have been translated into similar practices in classrooms with an emphasis on individualism and competition, on hierarchies. However, these features are neither supported nor practised by all research mathematicians. Nor are they to be found in the research practices of all mathematicians and could therefore legitimately, in my view, be said to be a distortion of the conditions under which effective mathematical learning communities are created and maintained. That such a distortion supports, even possibly enhances, unequal access to the discipline seems to me to be self-evident. One of the participants remarked: *"I have been getting very disillusioned with pure maths. and mathematicians; the arrogance within the subject never mind about anyone who isn't a mathematician."*

There is much, still, to find out through research about what makes mathematics interesting, exciting, appealing and accessible to different students and how such mathematics can be matched consistently to social requirements for useful mathematics. But students say that they do not like learning a subject that is disconnected from their lives or lacks utility in addressing questions that concern them. The mathematicians have provided an answer to this because, whether in pure mathematics, applied mathematics or statistics, the engagement and motivation, the creativity and commitment are part of what drives their learning enquiries.

APPENDIX A

Interview

The questions below are meant to give direction if we need it but not to be a requirement of how our conversation must develop.

About you:

1. Can we chart the historical trajectory of your becoming a mathematician?
2. From your experiences, what would you say you have learnt about mathematics?
3. What would you say you have learnt about yourself?
4. What would you say about how you come to know maths?
5. Is there anything you would like to say about your undergraduate or postgraduate experiences of coming to know maths?
6. How would you describe your experiences of research supervision and yourself as a supervisor of research students?
7. Of which mathematical community would you claim membership? Is that membership important and in what ways?
8. Do you have experience of collaborating on any research projects? Will you describe that experience and say what you have learnt from it or explain why you think collaboration is not helpful to you?

About how you come to know mathematics:

1. What do you now believe mathematics is? That is, how would you describe the focus of your work?
2. So what IS a mathematician, do you think?
3. Who are the mathematicians?
4. When you are acting as a mathematician, can you explain what you do, what choices you have, what leads you to make one choice rather than another?
5. Do you always know when you have come to know something new? How?
Have you been justified/unjustified in this confidence?
6. Has it always been like this?
7. Do you know whether a result will be considered important, interesting or rejected by your community?
8. Do you share their criteria? What are they?

9. Where do you find the problems on which you work and what makes them something which engages you?

Conduct:
Anything from the above list that you do not want to discuss we can delete before the interview begins. The interview will be audio-taped and I will also take handwritten notes. Typed notes of the interview will be returned to you afterwards so that you can agree their contents, amend, change or delete. These agreed notes will be the basis of the analysis which I do, backed up by the audiotapes where necessary. All information will be maintained as confidential. Copies of any subsequent publications will be offered to you

APPENDIX B

Study of Practising Mathematicians

Leone Burton
School of Education
University of Birmingham

Status: Prof/Reader/SL/L/Doctoral student

Personal questions

1. At what age did you 'know' that you wanted to 'be' a mathematician? (During the interview I would like to explore the historical route from that first understanding of what this mean to you, to your current understanding.)

2. Where did you study (undergraduate and postgraduate)?

3. Has there been anything in your post-graduate or post-doctoral experience which has influenced your images of mathematics?

4. Are you (a) an only child? (b) an eldest child? (c) placed in the family how? Please use B (brother), S (sister) and X for you and show oldest to youngest in line, e.g. BXSS.

5. Was anybody in your family connected with mathematics in any way?

6. Would you identify any particular influences in your journey towards becoming a mathematician?

APPENDIX C

The article reprinted here (with permission from the Journal for Research in Mathematics Education, copyright 2000 by the National Council of Teachers of Mathematics. All rights reserved) reported upon 53 papers provided by some of the interviewed mathematicians, and analysed discursively using the categories from the model derived in the original study. It was originally published in the *Journal for Research in Mathematics Education,* Vol.31 (4) 429-453, 2000.

Mathematicians Writing

Leone Burton, *University of Birmingham, UK*
Candia Morgan, *University of London, UK*

In this article we report on part of a study of the epistemological perspectives of practicing research mathematicians. We explore the identities that mathematicians present to the world in their writing and the ways in which they represent the nature of mathematical activity. Analysis of 53 published research papers reveals substantial variations in these aspects of mathematicians' writing. The interpretation of these variations is supported by extracts from interviews with the mathematicians. We discuss the implications for students and for novice researchers beginning to write about their mathematical activity.

Key Words: College/university; Connections in mathematics; Discourse analysis; Language and mathematics; Linguistics; Writing/communication.

In recent years, there has been increasing recognition in the mathematics education community of the social nature of mathematical activity and of the importance of communication within the practices of doing, teaching and learning mathematics (e.g., Boaler, 1997; Burton, 1999b; Resnick, Levine & Teasley, 1991; Steffe & Gale, 1995; Stephens, Waywood, Clarke & Izard, 1993; and with a particular focus on equity, Secada, Fennema & Adajian, 1995). Interest in mathematical language has gone beyond analysis of the mathematical symbol system and specialist vocabulary. The way in which language is used by teachers and students in classrooms is now a consideration as is its influence on the social identities and the views of mathematics that students construct (see, e.g., Pimm, 1984, 1987, on the implications of the use of *we;* Rowland, 1995, on the expression of uncertainty, and Gerofsky, 1996, on word problems). A considerable body of literature has appeared on writing to learn mathematics (see, e.g., Countryman, 1992). The language used in mathematical practices, both in and out of school, shapes the ways of being a mathematician and the conceptions of the nature of mathematical knowledge and learning that are possible within those practices.

Professional organizations have recognized the importance of engaging students in talking in the classroom, analyzing the nature of mathematical talk, and helping students to learn how to participate in oral mathematical discourse (Mathematical Association, 1987; National Council of Teachers of Mathematics, 1989). Although there is some interest in using writing as a means of learning (Connolly & Vilardi, 1989), less attention has been paid to the nature of written mathematical texts or to learning how to write mathematically (but see Morgan, 1998). Yet, writing plays a crucial role in many mathematical practices. The stakes involved in producing mathematical texts that are seen to be acceptable are often high, both for students who are likely to be assessed on the basis of their written work and for professional mathematicians whose status within the community and even job security may depend on the quality (and publishability) of their writing. In the educational context, students at many levels find writing difficult and may not communicate their mathematical thinking effectively (see MacNamara & Roper, 1992, on reports of investigative work in secondary school; Alibert & Thomas, 1991, on proofs at the undergraduate level). To help students (and more experienced mathematicians) at all levels to become effective mathematical writers, educators need to understand more about the kinds of writing they may be asked to produce and the forms that are seen as being appropriate for specific purposes. We, therefore, consider investigating the nature of writing in mathematical practices to be important. Our focus is the "natural language" within which the symbolic or special vocabulary and structures particular to mathematics are embedded. Although sometimes seen to be peripheral to the main mathematical content, natural language serves in the construction of the identities of the author and reader and of the epistemological and ontological assumptions underlying the writing.

Much writing about "the language of mathematics" treats it as if it were a unitary object that is the same in all circumstances. This approach is taken particularly when, as is often the case, research focuses on discrete features of language such as algebraic notation or special items of vocabulary. In whole texts, however, enormous diversity exists. Just as varying social practices may be labeled as *mathematics* (including academic mathematics, school mathematics, recreational mathematics, etc.), various genres of text may be called *mathematical* (e.g., research paper, textbook, examination question and answer, puzzle, etc.). The variety of types of writing used by mathematicians has been discussed by Mousley and Marks (1991), and, at a more general level, Richards (1991) identified at least four distinct "domains of discourse" and modes of argument associated with mathematics (research, inquiry, journal, and school math).

Most readers of this article can probably distinguish readily between extracts taken from, for example, an academic research paper, a primary school textbook, or the puzzle corner of a popular magazine. Not just the subject matter varies. The social functions that may be fulfilled by each of the texts also vary, including the construction of roles for both author and readers and of a relationship between them. Our current interest in the language of mathematical practices lies in understanding "what language is doing and being made to do by people in specific situations in order to make particular meanings" (Kress, 1993, p. 23). This understanding not only provides insight into "the social needs and the cultural values and meanings of its users" (p. 23) but also lies at the root of efforts to help mathematical writers to write more effectively, using mathematical language deliberately within their own specific situations to make the particular meanings that they intend. In this article, our focus is on the writing of research papers by professional mathematicians. By analyzing this writing, we hope to shed some light on the values and meanings of the practices in which these mathematicians are engaged. At the same time, however, the knowledge gained about the nature of the language used has potential to empower writers in this genre. Because the analytic tools used here (derived from Halliday's systemic-functional grammar, see Halliday, 1985; Morgan, 1996) may also be used for analyzing other mathematical genres, they may serve to support students learning to write mathematically in a variety of practices[1].

The fact that we can identify that a text belongs to a given mathematical genre--and that we can be surprised by a text that does not quite match our expectations--arises from the conventional use of particular linguistic structures. As novices try to gain acceptance and status within mathematical communities, they must produce texts that meet the conventional expectations of the gatekeepers within their fields. This requirement holds at all levels for students whose work is evaluated by their teachers and for new researchers seeking their first publications. Little attention has been paid, however, to the question of how novices are to learn what the conventions of the various mathematical genres are. Those advocating writing to learn mathematics appear generally to have assumed that students already know how to write or will pick up what they need to know through experience, even stating categorically, "We don't need to teach writing" (McIntosh, 1991, p. 423) in order to encourage mathematics teachers to try using writing in their classrooms. But, as Marks and Mousley pointed out, experience by itself may be too limited, both in quantity and in the variety encountered, to enable such natural development. "Writing is an unnatural act: it needs to be learned" (Reid as quoted in Marks & Mousley, 1990, p. 134). Marks & Mousley argued that, in order to develop mathematical literacy, students should be led "to development of an explicit understanding of the role of language in specific mathematical contexts" (p. 133). As Galbraith and Rijlaarsdam (1999) have pointed out "In order to learn to write pupils need to become aware of, and gain control of, the processes involved in producing text" (p. 102).

There is some debate in the domain of literacy education about whether students should be explicitly taught the characteristics of specific genres of writing (see, e.g., Reid, 1987). On the one hand, teaching the features of a genre is seen to be prescriptive and restrictive, curbing students' opportunities for creativity and self-expression (Dixon, 1987). On the other hand, it is seen as empowering, enabling students to participate in high-status forms of discourse. It is this latter perspective that we adopt. We believe that knowledge of the forms of language that are highly valued within mathematical discourses and of the effects that may be achieved by various linguistic choices would not just help writers to conform to conventional expectations but would also empower them to make informed choices to break the conventions in order to achieve deliberate effects, including to demonstrate creativity and to express their own personalities through their writing. Most of all, however, we adopt a pedagogic perspective on this debate because we believe that becoming critically aware, through analysis of text, of how it fashions and influences meaning can only enhance one's written communication.

The characteristics of genres are not fixed for all time but are always in the process of change, allowing possibilities for individual writers to participate in that change (Kress, 1993). Such possibilities depend, however, on knowledge of the conventional linguistic forms of the particular genre in question together with understanding of how these forms produce and are produced by the social relations, cultural values and meanings of the community. We refer to knowledge not just of the formal characteristics of the language (*langue*, to use Saussure's (1974) term) but, equally important, to knowledge of how the language is actually used (*parole*). In considering the genre of mathematics research papers, we examine the professional discourse of mathematicians, as represented in advice to writers published by professional organizations (e.g., Gillman, 1987; Knuth, Larrabee, & Roberts, 1989; Krantz, 1997), which presents a picture of correct or good practice, contrasting it with errors or what the authors consider to be infelicitous usage. This picture provides knowledge of (current) convention but, because of its normative style, does not empower its readers to view the conventions critically, to choose between alternatives, or to choose to adapt or reject the conventional forms in order to achieve particular effects. For such empowerment, an approach to mathematical language is needed that addresses what *can* be written and what can be achieved, rather than attempting to prescribe what *should* be done. In our analysis of the writing of mathematicians in this article, we are trying not to say that one way of writing is better than any other but to provide some understanding of the variety of ways of writing that is available within the research-paper genre and the effects that different choices may have on the social relationships, values, and meanings constructed.

THE STUDY

Purpose

In this article, we look at the language used in 53 published research papers provided to us as part of a broader study undertaken in 1997. We had two aims in analyzing the language of these papers. First we explored the mathematicians' perspectives as these might be represented in their writing--to identify what social positionings were possible and the extent to which different epistemological standpoints were apparent in the published work. Second, we attempted to describe some of the range of writing styles that may legitimately be available to professional mathematicians. In doing so, we had a particular interest in providing knowledge that might help students and novice mathematicians to make deliberate choices about the images of self and subject matter that they wish to construct in their writing.

Participants and Data Collection

Seventy research mathematicians, 35 women and 35 men, were interviewed for the broader study (Burton, 1999c). Because of their comparative rarity, the women were sought first through an appeal on electronic networks. Each woman's involvement depended upon both her willingness to be interviewed and her ability to find a male counterpart who also agreed to be interviewed. Of the female mathematicians, 17 were categorized as pure mathematicians, 10 were applied mathematicians, and 8 were statisticians. Of the male mathematicians, the numbers were 15, 13, and 7, respectively.

Procedure

The 53 papers we analyzed were obtained by a request, prior to the start of the interview, for a copy of a paper of the interviewee's choice. No constraints were put on the choice of papers submitted for this study. As may be seen in Table 1, a substantial number of these papers were written jointly with one or more other authors; we discuss the implications of co-authoring papers later in this article. These published papers were the focus of a discursive analysis utilizing epistemological perspectives raised within the interviews; when appropriate, our discussion is also informed by extracts from the interviews. We were interested in whether and how the variables of gender, status (as understood in the United Kingdom, i.e., lecturer, senior lecturer, reader, or professor) and the distinction among pure mathematics, applied mathematics, and statistics, might interact with the epistemological categories identified in the main study. Thus these categories are used, where appropriate, as descriptors in the body of the article. This practice has enabled us to formulate some conjectures to which we return later. The nature of the study, the choice of participants, and the complexity of the data are unsuitable for quantitative analysis, but at times we have presented quantitative summaries to inform the discussion and to raise particular issues.

Table 1
Characteristics of the Sample of Papers

	Single-authored (female/male)	Joint-authored (female/male)[a]	Totals
Pure	14(7/7)	9 (6/3)	23
Applied	8 (4/4)	12(5/7)	20
Statistics	2 (2/0)	8 (5/3)	10
Totals	24(13/11)	29(16/13)	53

[a]Only the gender of the participant in the study is included. The gender of co-authors not participating in the study has not been taken into account.

Framework for Analysis

In the case of academic papers submitted for publication, making a good impression on editors and reviewers depends on the appropriateness of the writing style as well as on the quality of the content of the paper. That writing style is important was pointed out by a male applied mathematics lecturer: "you learn that you certainly do have to write all the letters in exactly the form the editor wants or else you won't get to referee those papers, and they won't referee yours." In Anderson's (1988) studies of the reading of academic scientific papers he suggested that papers written in simpler language may be judged to be less valuable research and hence may be less likely to be accepted. This conclusion appears to contradict some of the advice on writing style offered to beginning academic mathematicians (e.g., Knuth et al., 1989; Steenrod, Halmos, Schiffer, & Dieudonné, 1973), which recommends that complex syntax should be avoided. It was also contradicted by the participants in the study, many of whom, when asked about criteria for publication, expressed strong feelings against obfuscating writing practices and claimed to place a high value on clarity and accessibility of writing. One said, "I would like my papers to be understood and well written and I would like the ideas to come across. The basic ideas to be clear, the insights, the understanding and the resolution, to come across." (Female Reader) Another pointed out, "Mathematical quality alone is not enough. I am very conscious of how the ideas are expressed. I like a story--a paper, a seminar are both stories. There is a beginning, a middle, and an end--so well written, clearly explained in a manner which is easy to follow and logical." (Male Lecturer)

But, like all judgmental categories, clarity is not absolute; it is relative to the context and, in particular, to the expectations and prior knowledge of readers. Moreover, there are some tensions between the ideal of clarity espoused in theory by mathematicians and the practice represented in their writing and in their relationships with editors and reviewers.

So how do beginners learn to write effectively? Is effective writing a matter of learning the expected norms and conforming to them, however much this may restrict what may be said or even conflict with the individual's self-image? Is there even a single set of norms? A critically aware writer (Clark & Ivanic, 1997; Fairclough, 1992) knows that there is always a choice about which form of language to use and that each choice is likely to induce a different effect. The challenge for the beginning mathematical writer is to know what choices are available, which norms can be broken, and what will be achieved by a particular choice.

To understand the norms and the choices available, we used features in the interview data to interrogate the 53 research papers from two major perspectives: the interpersonal and the ideational (Halliday, 1973). Halliday argued that every text, as well as performing an *ideational* function--saying something about the nature of the world--also performs an *interpersonal* function--saying who the author is and expressing the relationships between author and audience and between author and subject matter.[2] Halliday's functional framework and tools, derived from his functional grammar (Halliday, 1985; see Morgan, 1996, for a fuller discussion of the application of these tools to mathematical texts), helped us to formulate questions and explore them for each of the papers

Concerning the inter-personal function (*identity*), we tried to answer the following:

- How is the author's identity constructed as an *authority* in the field of research mathematics?
- To what extent and how is the author positioned as *a member of the community* of mathematicians?
- How is the author's relationship to the subject matter of mathematics constructed? In particular, to what extent does the author appear to claim ownership of the subject matter through *demarcation of knowledge or territory* within the field?

Concerning the ideational function (*focus*), we explored the following:

- What picture of the nature of mathematics and mathematical activity is presented? In particular, is the focus on mathematics as a *product of human mathematicians* or *an autonomous mathematical system*, or is it on *the processes of doing mathematics*?

APPLICATION OF THE FRAMEWORK TO EXAMINE RESEARCH PAPERS

Identity

Ask anyone what the characteristics of academic writing are, particularly in the sciences; the response is very likely to include the concept *impersonal*. Yet, according to Halliday's theory, every text says something about the author(s) and the reader(s) as well as about the subject matter. Our interest lies in challenging the assumed impersonality of mathematics texts and teasing out the ways that the personal and interpersonal materialize.

From our interview data, we found that academic writing was seen to be neutral and that its authority stemmed from the facts and arguments contained within it rather than from the personality or position of its author. This perception appeared to stem from "common knowledge", in the Edwards and Mercer (1987) sense of a common referential framework and a set of shared meanings established within a community through its joint interactions and discourse. Often, members of scientific communities make use of their established sets of meanings without explicit statement or conscious awareness, and the idea that writing should be impersonal coincides with ideas of scientific truth and objectivity. In mathematics in particular the apparent absence of the author from the text fits with positivist epistemologies in which the mathematician's role is subordinate to that of the mathematics itself. Davis and Hersh (1981) described the writing of their "ideal mathematician":

> "His writing follows an unbreakable convention: to conceal any sign that the author or the intended reader is a human being. It gives the impression that, from the stated definitions, the desired results follow infallibly by a purely mechanical procedure" (p. 36).

Studies of the language of academic scientific papers (e.g., Master, 1991), which consistently show a high proportion of passive constructions, provide evidence for this impersonal quality. For example, in the following extract from one of the papers, the agency of the author is obscured and a view of the subject matter as autonomous, independent of human activity, is suggested by using passive constructions[3]:

> "Now if the same procedure is followed for the solution of the terms proportional to $\underline{\Delta}^3\underline{E}$ in the perturbation quantities solutions can be obtained for $\underline{U}_{10}$, $\underline{V}_{10}$, $\underline{W}_{10}$, $\underline{T}_{10}$ and $\underline{\Psi}_{10}$. These are found to be identical to the linear solutions (4.12) with $\underline{f}$ replaced by the amplitude $\underline{f}_4$."

Here, as elsewhere, we are not suggesting that individual authors have made deliberate decisions about their forms of writing but they are conforming to cultural norms that produce the effects we are indicating. At the same time, those advising writers on style tend to suggest that authors should avoid the passive voice, perhaps because it is thought to be difficult to read. For example, notes on *Mathematical Writing* published by the Mathematical Association of America (MAA) advise. "The word 'we' is often used to avoid passive voice." They advised that an author use *we* in such cases only when meaning to include the reader, not to replace *I*: "Think of a dialog between author and reader". (Knuth et al., 1989, p. 2)

We noted that this advice was *not* taken in most of the papers we inspected. In single-authored papers *we* was frequently used to indicate the author alone; for example, a single-authored paper contained the sentence: "In section 2, we give the full compressible equations governing the flow of interest and the boundary conditions to be satisfied." In many other cases, for example, "We determine V_0 from the following relation...", it is unclear whether *we* is being used to refer to the author or whether the reader is being invited to participate in the determining. Interpreting the meaning of *we* in particular instances is often problematic. We (the authors and you, our readers) can only speculate about whether any individual author intended to invite her reader to participate in the actions described in the text or merely intended to report the author's own actions. Rotman (1988) pointed to a common knowledge assumption that readers must read mathematics actively, reperforming at least some of what has been done by the author in order to make sense of the text. Rotman distinguished the "thinker" and "scribbler" roles that are constructed for the reader by the use of imperatives in mathematical texts. Imperatives (*consider, integrate*), although unambiguously involving the reader, may or may not include the author as well (implicitly, *let us consider*). The use of *we*, necessarily including the author, is ambiguous as to its intention to include the reader, particularly in the light of its use in cases in which it cannot logically do so (as when used in the past tense to describe the author's processes). This apparent inclusivity may, however, have important effects on individual readers. Pimm (1987) reported his own rather negative response to a text that made extensive use of *we*:

> "The effect on me of reading this book was to emphasize that choices had been made, ostensibly on my behalf, without me being involved. The least that is required is my passive acquiescence in what follows. In accepting the provided goals and methods, I am persuaded to agree to the author's

attempts to absorb me into the action. Am I therefore responsible, in part, for what happens?" (pp. 72-73)

The sense of coercion or alienation that may result from a resistant reading of such texts seems to confirm the role that this indeterminate use of we plays in asserting the author's authority in relation to the reader.

The MAA advice continues: "In most technical writing, 'I' should be avoided, unless the author's persona is relevant."(Knuth et al., 1989, p. 2) The unwritten assumption here is that, in most cases, the author's persona is not relevant. This assumption is consistent with the common sense belief that the subject matter is neutral and that the use of such impersonal language reflects this neutrality. From our examination of the papers in our sample, we find that this rule, at least, is followed; I does not appear except in the most remarkable circumstances. In the 24 single-authored papers we examined, only 3 authors, 1 male and 2 females, used I. In no case of a multiple-authored paper, was I used as it was by Mason, Burton & Stacey (1982) to reflect the authors' "way of working, as well as the fusion that has taken place during writing" (p. xi). In our inspection of the 29 jointly authored papers, we carefully discriminated those authors who appeared to be using an agentic *we* (i.e., referring to themselves and their own actions) and those who seemed to be using we conventionally (as a means of avoiding the passive voice). In the former category were authors who, for example, appeared to refer to their own thought processes by asking questions such as "We may ask if consistency is necessary for convergence and for a certain class of difference schemes this is true." Such devices seem to introduce a degree of tentativeness and diminish the level of assertion. Agentic *we* was also used by those who explicitly described their own actions (*we* plus past tense): "In order to resolve the discrepancy we have carried out calculations on ... we have evaluated exactly ...." The conventional use (often characterized by *we* plus present tense) was more common, for example, "We apply this well-known procedure to graphical Gaussian model selection."

Among our participants there was a stark difference between those who wrote personally, for example, using the personal I and a modifier such as *just* ("Rather than producing an exhaustive list of work on quantum gravity using Regge calculus (which can be found in ...) I shall just mention a few typical applications in different categories.") or making meta-comments on their own writing ("The word 'generic' is used widely in mathematics; we use it in the precise sense of [cited authors], although our description will be informal.") contrasted with those who used the conventional, impersonal language in which mathematics is stereotypically presented. We noticed that some authors used a discursive approach in their writing, avoiding the conventional barriers between author and reader found in the majority of papers. For example, here an author shares with the reader an affective response to the subject matter: "Unfortunately there is not a very systematic definition of integrability. There is no shortage of examples, or indeed definitions, but the equivalence and general validity of the definitions is far from established ... My preferred definition ...." Very few authors used such personal or tentative language throughout as the general style but many authors introduced personal comment to an otherwise conventional presentation. Because such authors were more often, although not exclusively, male, we wondered whether women found it more difficult to use a personal or tentative style. The process of socialization into the discipline positions women as not *quite* belonging (see Burton, 1999a) and hence, perhaps, they are less able to risk rejecting the use of the language that they themselves see as representing competition and power.

The explicit expression of agency through use of the active voice with personal pronouns is not the only way in which the author's persona enters into a text. Relations among author, reader, and subject matter are expressed in the modality of a text, that is, in the "indications of the degree of likelihood, probability, weight or authority the speaker attaches to the utterance" (Hodge & Kress, 1993, p. 9). Modality may be expressed linguistically through use of modal auxiliary verbs (*must, will, could*, etc.), adverbs (*certainly, possibly*), or adjectives (e.g., I am *sure* that ...). The naive assumption might be that, because mathematics is about certainty, mathematical writing would always have the same absolute modality. That is, all verbs would be in the present tense and would be unmodified; no adjectives or adverbs would signal the author's degree of certainty about the result. Yet many of our participants remarked upon and deplored the frequent practice of signaling high modality in mathematics papers by the use of terms such as *clearly, it is obvious that* and *trivial.* In the interviews, the participants responded to a request to provide indicators as to what makes a paper, either their own or one they refereed, publishable. Many of the women focused, although of course not exclusively, upon style of writing--was the paper "reader friendly"--and drew attention to different styles and reasons for using language in particular ways, some picking out *clearly* for specific attention. One female lecturer said,

> *It ought to be understandable to people, but somehow in pure maths that doesn't happen. I try to do things in simple words and explain things fully, and [I] get quite cross when referees say you could chop this down; it's obvious. If something is obvious, there is no point in saying so, and it just irritates other people. Similarly clearly and all those [words]. It is the ideas being elegant, rather than the paper being short or concise.*

Another recognized some dangers implicit in clarity:

> *I think that scholarly articles are often written to impress people how clever you are. I try not to do that, I don't like it. If you explain something well it*

can sound quite simple but you don't want people to think it was easy, because it wasn't.

At the same time, many of our participants could, simultaneously, occupy quite conflicting philosophical positions during the interview, and we suppose, therefore, that their writing styles may not always be consistent with the content of their interviews. Speaking of the widespread use of words such as *clearly* and *it is obvious that* (words used to absolve authors from providing detail and to allow them to demonstrate their expertise), many of the participants said "No *clearlys*." Then they remembered that they had provided us with a copy of one of their published papers, and potential embarrassment, or at least awareness, would cross their faces and they would say something like "Well, if I use clearly ...."

From what we have seen, there is conflict and confusion among the conventional view of the impersonal nature of mathematical writing, the advice offered to writers, and the attitudes and practices of mathematicians. The beginning mathematician who needs to know what is possible and what is acceptable thus faces a problem.

In an attempt to make the available choices explicit, we turn now to look at the range and extent of differences in relation to the interpersonal functions performed by the forms of language used in our sample of papers. In the analysis that follows we have addressed three aspects of how the *identities* of the authors are constructed through their texts. These aspects reflect (a) the ways in which authors do or do not attempt to convince readers of their authority--positively (e.g., by assertion) or negatively (e.g., by tentativeness); (b) whether, and in what ways, they show identification with the community of which they are members; and (c) how they demarcate the intellectual territory and the knowledge to which they are laying claim. We deal with each of these aspects in turn.

Authority--positive and negative: The first aspect of Identity. One important aspect of the identities of authors as projected by their texts is the extent to which and the manner in which they claim to be authoritative within their community. If an author appears too tentative in his or her claims, less value might be placed on the results, whereas if one appears inappropriately self-assured, a reader might question the author's right to be so certain and even dismiss the work.

Terms such as *clearly* and *obvious* are relative to the individuals using them (implying this is obvious to me but may not be so obvious to you). Our interpretation that the use of such terms serves as a claim to authority on the part of the writer (implying this is clear to me and I do not need to explain it further because, if it is not clear to you, that is your fault not mine) is reinforced by the interview data, as demonstrated in some of the earlier quotes. We believe that, whatever the author's intent, the extent or absence of such words is one of the interpersonal aspects of the writing that will influence the ways in which the readers of the text will construct an image of the author and will consequently judge the worth of the text itself. Of course, there is not a simple causal relationship between the presence of particular words in a text and a reader's response; the way in which an individual reader responds to a particular text will depend on a complex interaction between the text itself and the resources brought to bear on it by the reader. These resources arise from the individual reader's social and cultural history, including the reader's self-perceived degree of expertise in relation to the subject matter, current positioning within the discourses in which the text is read (Kress, 1989), perception of himself or herself as an authority, and position in relation to the author(s). Expressing an appropriate and acceptable degree of authority is likely to be important for beginning mathematical writers in order that they may be accepted as (properly positioned) members of the community; Ward (1996) pointed out that "the author must consider the tactics to use in order to convince an audience of his or her truth claims" (p. 40). We felt, therefore, that the forms and extent of claims to authority were significant enough to warrant examination.

Indicators of authority claims from the use of modifiers such as *clearly, easily, of course, immediately obvious* and phrases such as *without loss of generality, it suffices to consider the case,* and *the last stage is trivial* all signal a gap in the argument (implying that there is no accompanying justification for the claim that generality is not lost) and hence signal to the readers that they should accept the author's claim. By using *well known*, the author claims the right to define the field and what is well known within it. Readers, in particular those defining the field of study differently from the author, might make different classifications of what is well known. Other indicators of authority claims include *for simplicity* and *for completeness*. The properties of simplicity and completeness are desirable within this discourse; thus the author is making a claim about the quality of her or his own work.

An important function of many of these phrases is to increase readability by removing unnecessary detail and to keep the paper to manageable length. Authors, however, in making decisions about what is unnecessary, explicitly exercise their authority in relation to both subject matter and reader.

Authority claims were sometimes made by denigrating the work of others: "The paper contains a great deal of notation, which, to our way of thinking, obscures the ideas. On occasion the author seems to assume the main point he is trying to prove." In this way, the author of this comment exercises his or her own right to determine what is of value in the field.

Although in the interview some participants expressed dislike of such authority claims and asserted that they would not make extensive use of them, remarkably few texts among those we examined completely avoided such claims. Of those that did avoid authority claims, four (3 females and 1 male) were extremely impersonal texts with few indicators in any of our categories related to the construction of the identity of the authors. Three others (all by male authors) had unusually high numbers of indicators of negative authority, suggesting a possibly deliberate attempt to relate in a less authoritative way toward their readers. (Table 2 summarizes the distribution of numbers of positive and negative authority indicators.)

Table 2

Extent of Indicators of Positive and Negative Authority in Each Paper

	Number of indicators of positive authority[a]				
	0	1	2-10	11-20	>20
Females	3	3	18	4	0
Males	4	1	11	5	4
All	7	4	29	9	4
	Number of indicators of negative authority[b]				
	0	1-3	4-6	>6	
Females	11	9	9	1	
Males	4	9	5	5	
All	15	18	14	6	

[a]median = 4
[b]median = 2

We analyzed the submitted papers to determine whether some authors either failed to exert such authority or even explicitly admitted a lack of knowledge or certainty. Signs of such negative authority included not only absence of the positive authoritative signs described above but also modifiers such as *difficult* or *not obvious* and comments such as "the integral in this expression can only be calculated numerically so there is no 'quick and easy' method of generation available;" all of which indicate that the author has encountered difficulties. The author is thus vulnerable to a reader who does not accept that difficulties exist. Comments such as "we think but have not yet shown," "plausible though unproved assumption," "the quantum stability is an open question," "it has to be admitted that our reanalyses were idealized in that ...," and "this identity is too crude a tool" expose the authors' failure to meet completely the standards demanded by the community and thus constitute a risk to their status. Sometimes authority was transferred to others by saying, for example "It is reassuring that the results presented here are similar to those [of X] ...," a statement that could result in a diminution of status for the author.

There were fewer indicators of negative authority than of positive authority claims, as can be seen from the distribution also shown in Table 2. There were, however, nine females and six males in whose papers we found more negative than positive indicators or an equal number of each. At the other end of the scale, substantially more female than male authors avoided taking the risk of including any explicit indicators of lack of authority. Although we have described negative authority in terms of vulnerability and risk for the author, the seriousness of this risk is, of course, relative to the author's preexisting status within the community. For example, a very eminent male professor of pure mathematics provided a paper that contained unusually high numbers of indicators of both positive (59) and negative (7) authority. His paper also appeared extreme in other categories discussed below, suggesting strong orientation toward the community (8 references to the community and 18 citations of other authors) and focusing extensively both on mathematics as an autonomous system (50 indicators) and on the human processes involved in mathematical activity (46 indicators). A high-status author can afford to state that a piece of mathematics is difficult in the confident expectation that others will not challenge this judgment (and may even question their own understanding of the mathematics if they do not perceive the difficulty themselves). For low-status authors, such statements would be likely to reinforce any existing doubts about the value of their work; such authors, therefore, may feel themselves better advised to construct a positive authoritative identity in their writing. As a female Senior Research Officer said,

> *One of my collaborators likes to make things seem as sophisticated and complicated as possible and I think that is a problem for us. If I write the paper, our result looks like a simple result and if he writes it, it looks like something else completely. I remember giving a seminar once and being told afterwards that I had made it look far too easy.*

The texts that we have examined in this article are located within the discourse Richards (1991) called "journal math", using in most cases "reconstructed logic" (p. 16). Solomon and O'Neill (1998) argued that this discourse, which predominately makes use of the "timeless" present tense and logical cohesive devices (e.g., *hence* or *so that*), is constitutive of mathematics itself. We agree that such linguistic forms are essential for expressing many mathematical meanings but deny a necessity that logical, rather than temporal, relations make a narrative form unavailable. On the contrary, we believe that a mathematical text can be seen as a narrative, albeit told within the different conventions of that genre. With McLaren (1995), we see narratives as providing "the discursive vehicles for transforming the burden of knowing to the act of telling" (p. 92). However, we also feel that to exclude the personal narrative of the "logic of discovery" (Richards, 1991, p. 15) neglects other important aspects of mathematics. Those papers in our sample that made the author's role in the production of the mathematics clear or constructed an active role for the reader were not any less mathematical but indeed presented a broader view of mathematics, including the human mathematician in the picture. To include the activity of doing mathematics as well as the results and the reconstructed logic may be one way of making such texts more accessible to both writers and readers. But doing so also demonstrates that "science is not the supreme way of knowing simply because of its philosophical, ideational, or rhetorical supremacy, but because it is a powerful associational network containing a strong and expansive web of heterogeneous actants" (Ward, 1996, p. 138). Having said this, we would not wish to argue that there is one best or correct way of writing mathematical research papers. To do so would merely be to substitute one form of authority for another. Rather, we would like to see greater tolerance of diversity and, as important, awareness of the variety that already exists and is available for writers to use to express their own chosen mathematical and personal meanings.

Community membership: The second aspect of Identity. In the research project of which this forms part many of the participants spoke in similar positive terms to those of the lecturer who said, "I am an applied mathematician - a fluid dynamicist. My community provides me with colleagues to talk with, and work with, a place to find new ideas." We were, therefore, interested to investigate how their writing demonstrated this identification and relationship with the community. One feature that was striking in the interview data was the change in recent years from individualized to collaborative research (see Burton, 1999c). One participant referred to this change as a cultural shift in the discipline. All except 4 of the 70 participants claimed to be doing some collaborative work so we examined the effect of collaboration in the fifty-three papers we analyzed and found that 24 of the papers had one author, 19 two authors, and 3 four authors.

Gender did not appear to be a differentiator (see Table 1); approximately 55% of both female and male participants gave us a jointly authored paper. Of course, the choices of article were not random and did not necessarily reflect the authors' current work or thinking about how they preferred to work. Indeed, we believe it is likely that jointly authored papers are underrepresented in the sample. Three factors may have led to such underrepresentation. First, the interviews were individual so the participants may have assumed that it would be preferred or better, in some way, to offer a single-authored paper. Second, the content of the interviews dealt with epistemological questions about their mathematical thinking, how they understood mathematics, how they organized their work, the role of intuition and insight, and so on; again, these questions could have influenced participants to assume a single-authored paper would be preferred. Finally, a few participants gave us more than one paper, reflecting different ways of working, different areas, changes from one period to another. In these cases, we chose a single-authored paper to contribute to this analysis.

Having identified Community Membership as an area of interest, we examined the papers with respect to the ways in which the authors positioned themselves in relation to the wider community of mathematicians. The first indicator was a simple count of the number of times that the author(s) made reference to the work of others (other-citations) and the number of times they cited their own work (or cited a coauthor). The other-citations ranged from 0 to 58, and self-citation ranged from 0 to 10, but the lengths of papers also varied considerably as did their purposes, which made comparisons difficult. Citations, of course, serve a wide range of functions (see, e.g., Cronin, 1984), but we have used them in this study simply as indicators that authors have located themselves in some (undetermined) way within a community of other authors. The highest number of self-citations (10) was made by a female. The highest number of other-citations (58) was made by a male.

We also explored the texts to find other instances in which the authors referred to the community of mathematicians to which they belonged and to see how frequently they did this. Some remarks were nonspecific, referring in general terms to work ongoing: "The rich microstructure of cosmic strings is starting to receive considerable attention." In other remarks, authors cited particular persons, for example, "since the pioneering work of [X]," some comments located the authors in relation to specific contributors to the field and to a broader community: "The problem is a difficult one and there have been numerous papers written on the subject including an excellent review by [Z]." As can be seen in Table 3, the number of such references ranged from 0 to 22. Gender did not appear to be a differentiator.

Table 3
Numbers of references to the community

	References				
	0	1-4	5-9	10-15	>15
Participants	2	31	15	2	2*

*One of these participants made 17 references to the community and the other made 22.

The majority of our authors are positioning themselves in general with respect to the community or citing a small number of appropriate references, but there is considerable variation in practice. We are not convinced that counting citations in this way has given us a clear indication of community identification or of the conventions of academic writing in mathematics. The content of the interviews was convincingly community related, but the published writing did not clearly reflect this fact.

Demarcation of territory and of knowledge: The third aspect of Identity. The worth of a piece of research is likely to be judged not only by its internal validity but also by the extent of its contribution to the field of knowledge of which it claims to be a part. Indeed, the mathematicians interviewed were quick to invoke a number of words to describe categories of contributions, from *significant* or *important* to *interesting* and, dismissively, *trivial*. These categories appeared to form a hierarchy in their thinking, of which the pinnacle was a *significant* contribution. Work that is derivative or that reviewers or editors judge to be uninteresting or trivial is unlikely to be published. A lecturer in applied mathematics said

> *To decide whether work is important, I would ask what sort of application, how significant is the application, how significant is the result. Does it back up something someone else has done? Does it tie in very well with the experiments which could validate your model or method? You could have some experiments that had been done elsewhere; you could do some nice maths matched up with it, which is satisfying in itself, but if it isn't going to help in pushing forward design, it might be nice in itself but it isn't going to be so significant to the community.*

The criteria in other areas of mathematics may be slightly different from those expressed by this lecturer, particularly in pure mathematics in which experiments are unlikely to be a means of validation. Whatever the field, however, authors are likely to be concerned with demonstrating the place their work may occupy within it. As well as establishing one's authority to speak within the community of mathematicians, one of the interpersonal functions that a research paper may thus fulfill is the demarcation of an author's claim to be entitled to speak about a particular area of knowledge. On examining the papers, we found that we could distinguish two types of such demarcation: the demarcation of territory and the demarcation of knowledge.

In demarcating territory, authors separated their own work from that of others in the field, identifying gaps in the community's knowledge and stating what remains to be investigated. In the papers we examined, authors not only located and made a claim for the unique contribution made by a paper (e.g., "One way in which our treatment differs from all previous treatments, is that we do not assume we start with an embedded fundamental domain.") but also laid a claim to areas of work not, as yet, published ("That is already a substantial programme of work ...which we hope to develop in future publications"). In some cases, claim was made on areas yet to be investigated, thus establishing a larger territory (and hence, perhaps, greater status for the author) not directly justified by the contents of the current research paper or ongoing research: "As an extension to this work we intend to investigate nuclear vibrational motion ... Future calculations are planned to employ a multistate approximation ... We also plan to carry out *ab initio* studies of photoionization cross sections...." By making such statements, authors not only claim status for their current work but also attempt to ensure that they will be able to make further claims in the future (and that other workers in the field will find it necessary to cite their contributions).

As well as demarcating areas of their field of mathematics, authors also claimed status for the knowledge produced within the field and, in particular, for the value of their own results. In some cases, simple statements of ownership of results achieved this demarcation of knowledge, for example, "We establish a general multi-transversality theorem...." In other cases, more ambitious claims for the value (in terms of significance or interest) of these results were involved: "The algorithm presented here could lead to significant saving of time and complication."

The extent to which the papers included such demarcation of territory or knowledge varied substantially although only one paper contained none at all. Interestingly, we found little relationship between the extent of authority claims and the extent of demarcation in individual papers, suggesting that establishing the author's personal authority in relation to her or his readers and the author's right to recognition as the owner of an area of mathematics are two separate aspects of the interpersonal function of the text.

Focus

In the interviews we found great variability among the participants about the nature of the beast, mathematics. One mathematician said, "The only thing mathematicians can do is tell a good story, but those stories do uncover mathematical truths--mathematical truths are discovered." Another explained, "I see mathematics as a way of describing physical things using formulae that work, modeling the real world, and having a set of rules, a language, which allows you to do that."

Although both these mathematicians talked in terms of the communicative aspect of mathematics, the first, a pure mathematician, is telling a story, whereas the second, an applied mathematician, is describing aspects of the world. Of course there are many cultural differences between pure and applied mathematicians (see Davis & Hersh, 1981 and Burton, 1999c), but, from the perspective of mathematical writing, we were particularly taken by the difference between assertions of certainty made by most of those who were interviewed (58 of the 70) and the comments by a small number of participants (mostly statisticians), who deliberately contrasted their knowing with certainty:

> *If I know something, it means I understand how a particular system is described mathematically and what the mathematical implications are...Sometimes you never know if you know. You can do it by contradiction, by finding fault. You can never know it is right. You always have to live with uncertainty. (Professor of statistics)*

Jane Hutton (1995) also affirmed that "statisticians ... are concerned with variation and error, and the resulting near impossibility of knowing anything with certainty" (p. 254).

We found another distinction between those in pure and those in applied mathematics. Applied mathematicians tended to emphasize utility: "You can do things with it; you can model real things; you can make predictions; you can compare experiments." They were mainly concerned about "describing physical things using formulae that work, modeling the real world." But even this very practical approach did not preclude this mathematician from acknowledging the sociocultural basis by "having a set of rules, a language, which allows you to do that." A pure mathematics lecturer gave an interesting example of a very different interpretation of the nature of mathematics:

> *The definition, theorem, proof style is sometimes necessary to the health of mathematics, but it can be overprescriptive. People think that is what maths is whereas I think it is about filling in gaps, making the map. Maths isn't what ends up on the page. Maths is what happens in your head. I don't think maths is about proving theorems. It is one constituent, but maths is about mapping abstract ideas in your head and understanding how things relate.*

In the light of these differing images of the nature of mathematical activity, we were interested to investigate the focus of the papers that we were analyzing. Within this category we explored two aspects of the writing: the representation of the source of new mathematical knowledge (whether human or from within the system of mathematics itself) and the extent to which the human processes of doing mathematics (as opposed to presenting the results of this activity) were apparent in the texts. The highly stylized way in which mathematical papers are often written cannot necessarily be reinterpreted as intent or as representative of particular positions. Indeed, as earlier extracts from the interviews have shown, experienced authors are aware of the expectations of their audience and of the ways in which they may need to adapt their writing to match these expectations. So although the writing of naive beginning writers might represent their beliefs more or less transparently, we cannot assume that the same will be true of more experienced authors.

Mathematics-Human product or autonomous system? The first aspect of Focus. We looked for differences between authors who made the mathematics the object of their attention by focusing on the products (proofs, results, etc.) of their own activity, using phrases such as "our claim is proved" and those who appeared to give the mathematics a "life" of its own (indicated by the use of words that are generally associated with animate beings: "The prolonged good behaviour of the Hartree-Fock type wavefunction ...is probably due to...") and by the extensive use of clauses in which mathematical objects themselves acted as the subjects of material activity (as opposed to being operated upon by a human mathematician, for example, "The functors H and K induce a duality ...").

In many of the papers, the mathematical objects were themselves the actors producing the results that were being described by the authors. Frequency of use of this style varied from not at all, to 50 such examples. We looked closely at those papers in which authors used no such examples and those that had more than 20 examples to try and demarcate differences. Two female statisticians avoided the use of this style entirely, choosing to focus upon humans as responsible for the mathematics. Both made references to mathematicians as producers and to the processes whereby the mathematics was produced (see next section). At the other end of the scale, all the papers with more than 20 examples of mathematical objects presented as actors came from male participants. There may, however, be a relationship between the extensive use of this style and the field of mathematics. In particular, the structure of the discourse of some areas of pure mathematics seems to lend itself to a focus on the autonomous activity of mathematical objects. A further result of the removal of humans from the writing is that the style then tends to become obscure and abstract; we have already referred to this as one way in which an author can seek to establish her or his authority (and see Davis & Hersh, 1981).

In some cases, authors may have presented the mathematics as a human product as a matter of conventional presentation, including the convention of marking the completion of a proof with a statement to that effect. However, in other cases of papers with this focus, the style was not solely conventional but was tempered with process comments--indicators of a focus on human activity as well as human product. Thus, alongside comments such as "We can obtain an algebraic equation", we found "We attempt to prove stability". A strong focus on mathematicians as producers sometimes accompanied strong acknowledgement of the community, leading us to wonder if these are connected, but our data are insufficient to test this hypothesis.

The processes of doing mathematics: The second aspect of focus. A further aspect of interest to us was the extent to which the focus of a paper was on the processes by which the authors arrived at the end product of theorems and proofs rather than solely on the products themselves. We looked for indicators of a focus on the human processes of doing mathematics such as expressions of opinion or feelings, conjectures, questions, explanations for decisions (e.g. "This is a purely analytic proof which gives little insight into why ..." or "The motivation for this paper was an attempt to build models").

The number of explicit mentions of human processes involved in doing mathematics ranged from 0 to 46 examples in a single paper. We might expect a complete absence of process comment to indicate a very conventionally written paper with a strong focus upon mathematicians as producers, but this was not necessarily the case; some of these papers also used the language of negative authority, acknowledging the community but focusing on the knowledge product rather than the process of its production. Authors who used a high number of examples describing processes often appeared to be more strongly assertive of these processes than they were of their claims to knowledge. We particularly noted that some authors who made frequent reference to process also asked questions of the reader, (e.g. "How do we know that..? And, if so, how can we find ...?") thereby involving them in the paper in what we felt to be a very user-friendly way. Because of the important role that humor can play in introducing the humanity of author and reader as they share the joke we appreciated the ways in which two authors used humor, "The noose is a good place from which to hang these ideas"; "The details of this argument are thankfully hidden under the very general apparatus of G-structures."

CONCLUSIONS

During her interview, a female Senior Lecturer was particularly critical of mathematical writing:

> *Mathematics is full of hidden assumptions. You have to read a paper at least twice to get the real agenda, and you have to go through a process of deduction to find out what on earth it is. Is it that the people who are doing it are blithely unconscious of the wider process or is it a process of mystification? I think sometimes they are only semi-conscious of what they are doing and why.*

Hersh (1989) also blamed poor mathematical communication on authors' failure to have learned how to write in prose. In spite of widespread agreement within the mathematics community that much mathematical writing at all levels is obscure and difficult to read, remarkably little guidance is available for beginning or experienced writers. This lack of guidance contrasts with the situation in science and technology in which a substantial number of publications offers guidance on writing, apparently reflecting a recognition that many of those involved in these areas are likely not only to have underdeveloped skills in this area but also, at some point in their careers, to be involved in writing reports for nonspecialists, with the consequent need to focus on the clear communication of key ideas. The Mathematical Association of America has shown some interest in publishing guidance on mathematical writing (see Gillman, 1987; Knuth, et al., 1989, Krantz, 1997), but the training of mathematicians does not appear to include any systematic attention to the development of writing skills. Moreover, the guidance that is available is largely algorithmic, sets of rules to be followed rather than ideas for developing critical awareness of the different effects of various forms and structures of language. As the mathematician quoted above suspects, many mathematical writers are likely to have little awareness of how their readers may respond to their writing. In particular, given the rule-based nature of the guidance that is available, authors are unlikely to be aware of the possibilities for variation and flexibility in style that we have found in our examination of these papers, despite the writings of authors such as Steven Ward, who points out that "rhetoric and literary invention are key communal ingredients for doing and conveying science...if scientists and other knowledge producers can see how rhetoric and symbolization work in the writing process, they will be able to write better texts" (1996, p. 40).

However, apparently not all mathematicians are happy about the ways in which mathematics is presented publicly. A female Reader asked that papers should be "well written". She continued

> *I get annoyed with some of my collaborators and a lot of the papers I am sent which are definition, theorem, lemma, proof. That seems to me to be appallingly bad. It is the sort of thing that no one is ever going to want to read. I think it is important to grab the reader from the opening sentence. Not "Let A be a class of algebras such that ...." Change it to "This paper opens a new chapter in duality theory."*

We believe that the dominance of a style of writing that she, and many other participants, abhorred is probably the result of a combination of three factors: desire for feelings of safety, the quantity of published papers providing conventionally written models, and a nonreflective approach to writing. Yet there were substantial differences among the papers the participants gave us. "Differences between writers reflect not just differences in their cognitive skills but also the extent to which they share the goals, and understand the conventions, of the discourse community they are addressing" (Galbraith & Rijlaarsdam, 1999, p. 98).

We hope that we have demonstrated the diversity in the variables we have examined and the lack of universal conformity to commonly acknowledged conventions. Other strong influences on writing-style choices may include the authors' epistemological stances on mathematics as well as their beliefs about the expectations of their readers. Nonetheless, our analyses of the various aspects of identity demonstrate that a research paper does more than report research. We have shown how authors, through their writing, convey very particular perspectives on themselves, as well as on their mathematics, and on how they build their identities with respect to their communities.

If some mathematicians want to move away from obfuscation, mystification and put-down toward clarity and straight-forwardness without oversimplification, the knowledge that some authors attempt to follow these principles would provide them support. However, editors of journals and their chosen reviewers also have a role to play in emphasizing certain principles that, at present, do not seem to influence either those who submit papers for publication or their peers who review. Textbooks also play a role in formalizing the mathematical language that is accepted. During the past 10-15 years, in some social science writing, noticeable shifts have occurred away from the conventional, third person, objective writing toward a style that incorporates the I of the author and links the author's beliefs and values more clearly to the chosen structure and outcomes of the research. Some social science journals provide authors and reviewers with guidance about, for example, the use of non-gender-specific language. Providing similar guidance to mathematicians could be an important step in raising awareness of the power of language. We hope that this article may contribute to such a process.

For research mathematicians, writing (and, more important, being published) is a critical activity. Although it may be seen as secondary to the act of doing mathematics itself, through their writing individual mathematicians establish their identities within the academic community and secure employment. Common knowledge about the nature and process of mathematical writing (as represented, for example, in the MAA guidance mentioned above) suggests that there is only one standard way of presenting one's research. Yet our examination of this relatively limited sample of published papers has shown considerable variation and, in a few cases, wholesale flouting of the recognized conventions. The conventions of mathematical writing are neither necessary nor natural consequences of the nature of the subject matter; they are rather "the product of current relations of power and discourse practices" (Clark & Ivanic, 1997, p. 14) within the community. Knowing the conventions and being able to use them may be one step toward establishing one's position; knowing how they work and how and when to break them to achieve a particular effect is, however, an important way to express and establish a more powerful position.

We see teaching and learning mathematics not as just filling students' heads with facts and skills but as inducting them into mathematical communities. An important part of learning to be mathematical, whether in the primary school or in the university, is learning to take part in the discourses of mathematics, becoming both a consumer and a producer of texts that are recognized as legitimately mathematical within one's community. Current practice in the training of mathematicians and in mathematics education more generally does not explicitly involve teaching and learning about mathematical writing. The novice may learn through using the existing models of published writing, through an apprenticeship of collaboration with more experienced writers, or through the often harsh process of peer review. None of these methods is designed to help learners to acquire the kind of knowledge about language that might enable them to be aware of what they might achieve by choosing to write in different ways. Moreover, although the rules governing what is and what is not considered acceptable writing within the community remain implicit, access is likely to be more difficult for those from social groupings (of class, gender, or ethnic group) whose linguistic resources do not match those of the dominant group. At the level of school mathematics, one of us has argued (Morgan, 1998) that teaching should address mathematical writing explicitly to make successful participation in mathematical practices more accessible for all students; the same need exists at more advanced levels. For such explicit attention to writing to be possible, teachers and students need to develop a mathematical-linguistic vocabulary that equips them with tools of thinking and speaking about the various forms of language available to them so that they can make informed, critical judgments about which forms to use. We see this acquisition of mathematical-linguistic vocabulary as being a necessary and crucial component of the educative process. The tools that we have offered here and used in the analysis of research papers provide a starting point for the development of such a vocabulary for those learning to write in this genre. A more general set of tools is suggested by Morgan (1996); further work needs to be done, however, to characterize other mathematical genres and the choices available to writers within them, particularly in those genres in which producing appropriate text has important consequences for learners and novices. Meanwhile, mathematicians, teachers, and learners can embark upon the process of learning to read and write mathematics critically.

Writing, for both students and researchers, is not just about communicating mathematical subject matter. It is also about communicating with individual readers, including powerful gatekeepers such as examiners, reviewers, and editors. The writer needs to know how to write in ways that are likely to convince such readers that she or he has the authority to write on this topic, that the subject matter is important enough to be interesting, that it is worth paying attention to what is being said. This need for knowledge about language is as great for learners of mathematics as it is for practitioners. Where, as in the United Kingdom, new developments in assessment demand that students present mathematical projects, extended essays, investigations, and so on, the students' experience of considering how and in what form to convey their meaning becomes even more important than in systems whose assessment relies on multiple choice tests or tasks involving short responses.

We have seen that authors can write effectively and powerfully without subsuming their identities within the conventions. Indeed, the papers of some mathematicians with the highest status within the community used some of the least conventional language. In this exploratory study we have identified linguistic means for achieving various types of authority, significance, interest, and so on. But characterizing the various forms used by mathematical writers requires further research. This study could be a starting point for work with novice (and, indeed, experienced) researchers to develop their critical linguistic awareness--their knowledge of the forms of language that are available to them and their abilities to make effective choices among them. We hope that our study may help join potential mathematicians, learners of mathematics, to practitioners, those who are researching in the community of mathematicians.

[1] See Morgan (1998) for an example of analysis of secondary school students' writing of reports of mathematical investigations and discussion of how this may be used to support learning to write in this genre.
[2] Halliday also identified a textual function--making the text into a meaningful message--that we do not consider in this article.
[3] Note the effect on the reader of our own choice of the passive voice.

Acknowledgements

We are particularly grateful to our participants for their generosity of time and for providing papers and to the anonymous referees' helpful comments on this article.

APPENDIX D

APPENDIX D

Tree Diagram of NUDIST categories

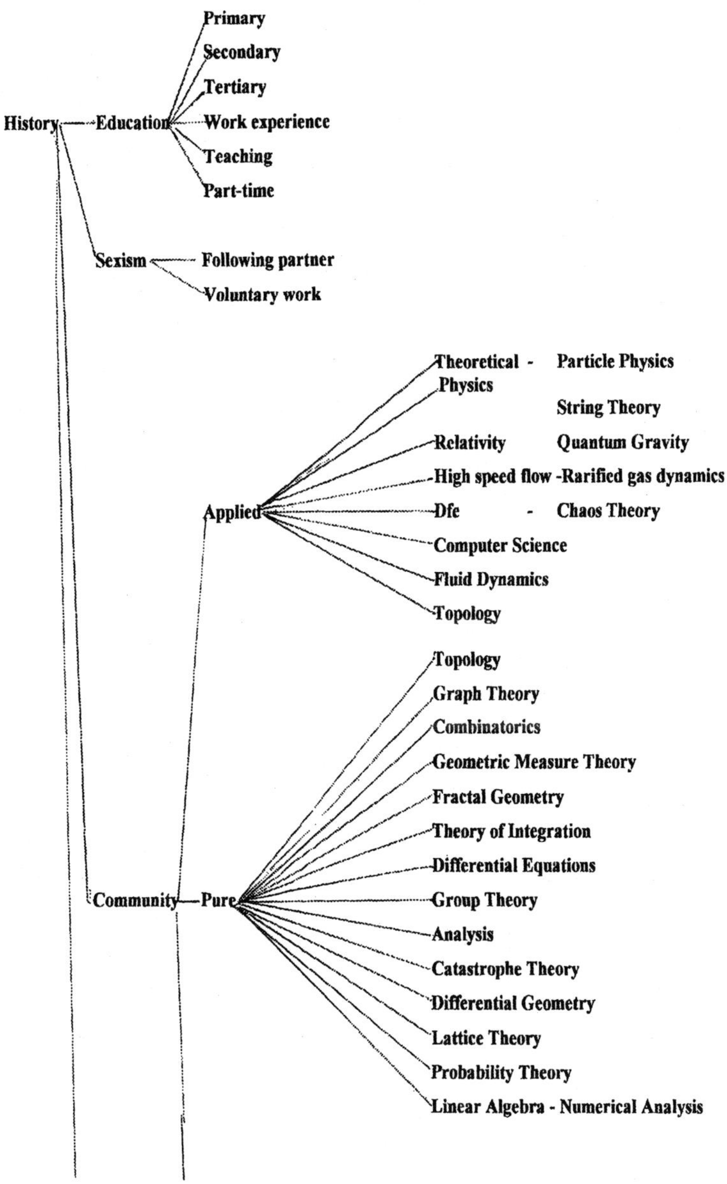

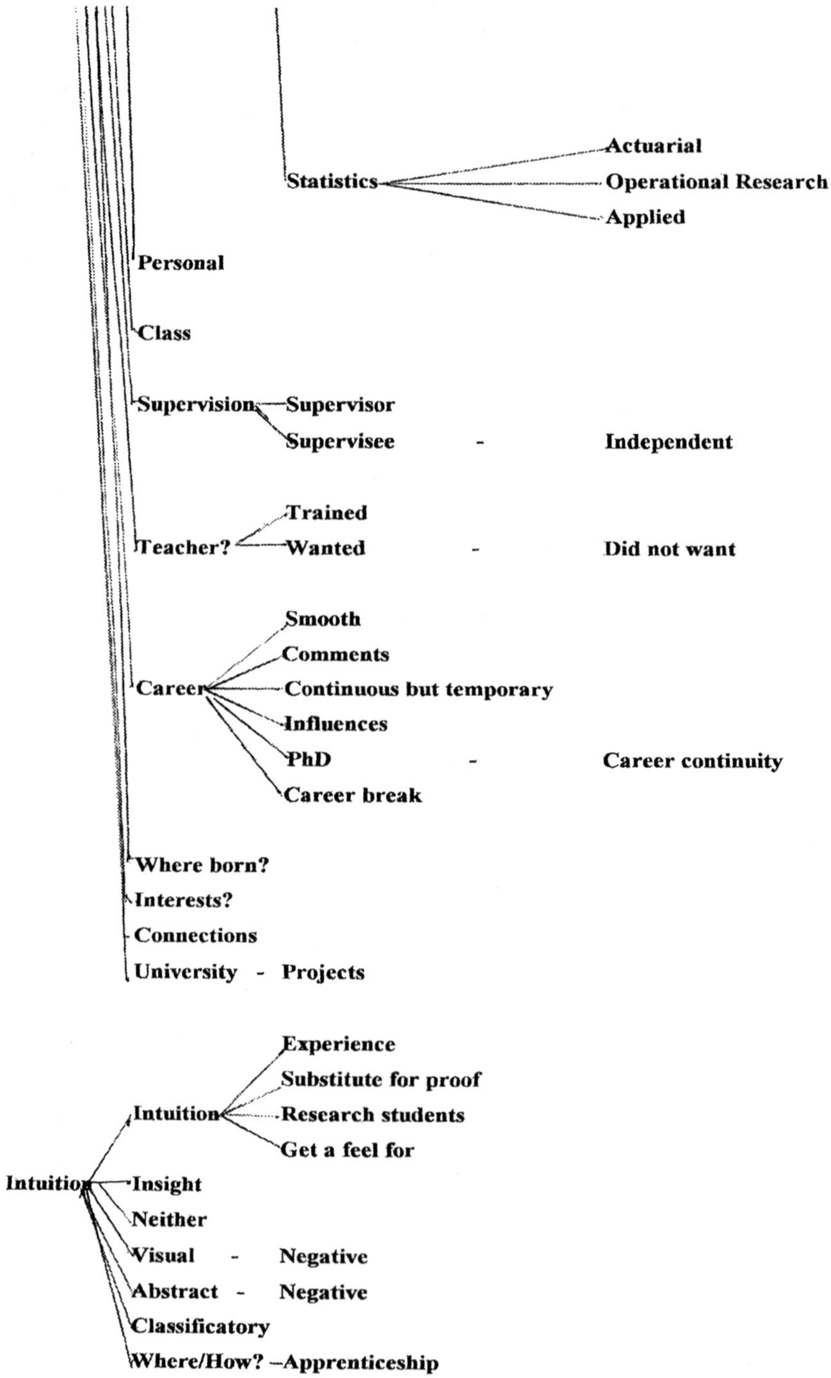
Statistics
Actuarial
Operational Research
Applied
Personal
Class
Supervision
Supervisor
Supervisee - Independent
Teacher?
Trained
Wanted - Did not want
Career
Smooth
Comments
Continuous but temporary
Influences
PhD - Career continuity
Career break
Where born?
Interests?
Connections
University - Projects
Intuition
Intuition
Experience
Substitute for proof
Research students
Get a feel for
Insight
Neither
Visual - Negative
Abstract - Negative
Classificatory
Where/How? –Apprenticeship

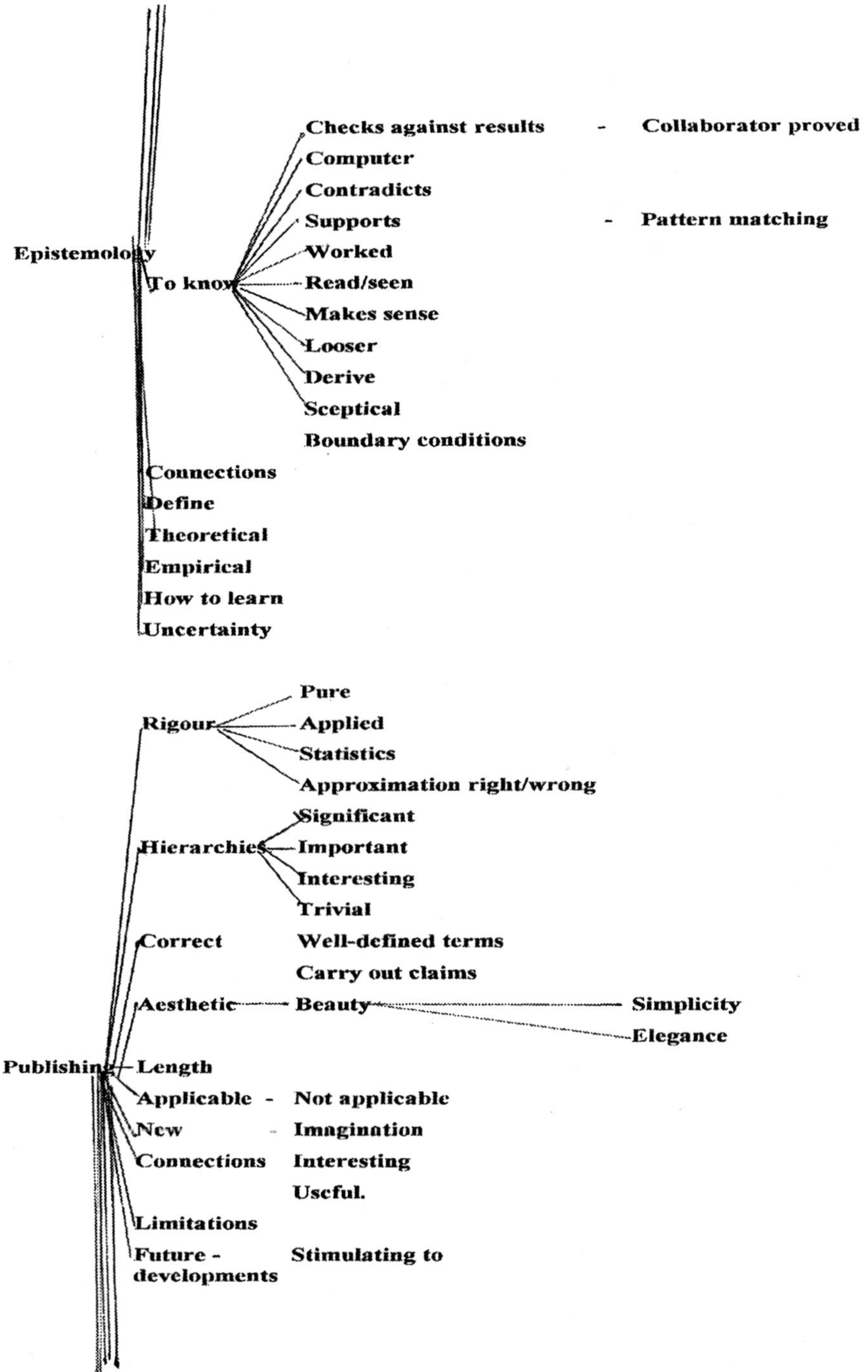
Epistemology
To know
Checks against results - Collaborator proved
Computer
Contradicts
Supports - Pattern matching
Worked
Read/seen
Makes sense
Looser
Derive
Sceptical
Boundary conditions
Connections
Define
Theoretical
Empirical
How to learn
Uncertainty
Publishing
Rigour
Pure
Applied
Statistics
Approximation right/wrong
Hierarchies
Significant
Important
Interesting
Trivial
Correct
Well-defined terms
Carry out claims
Aesthetic
Beauty
Simplicity
Elegance
Length
Applicable - Not applicable
New - Imagination
Connections
Interesting
Useful.
Limitations
Future - developments
Stimulating to

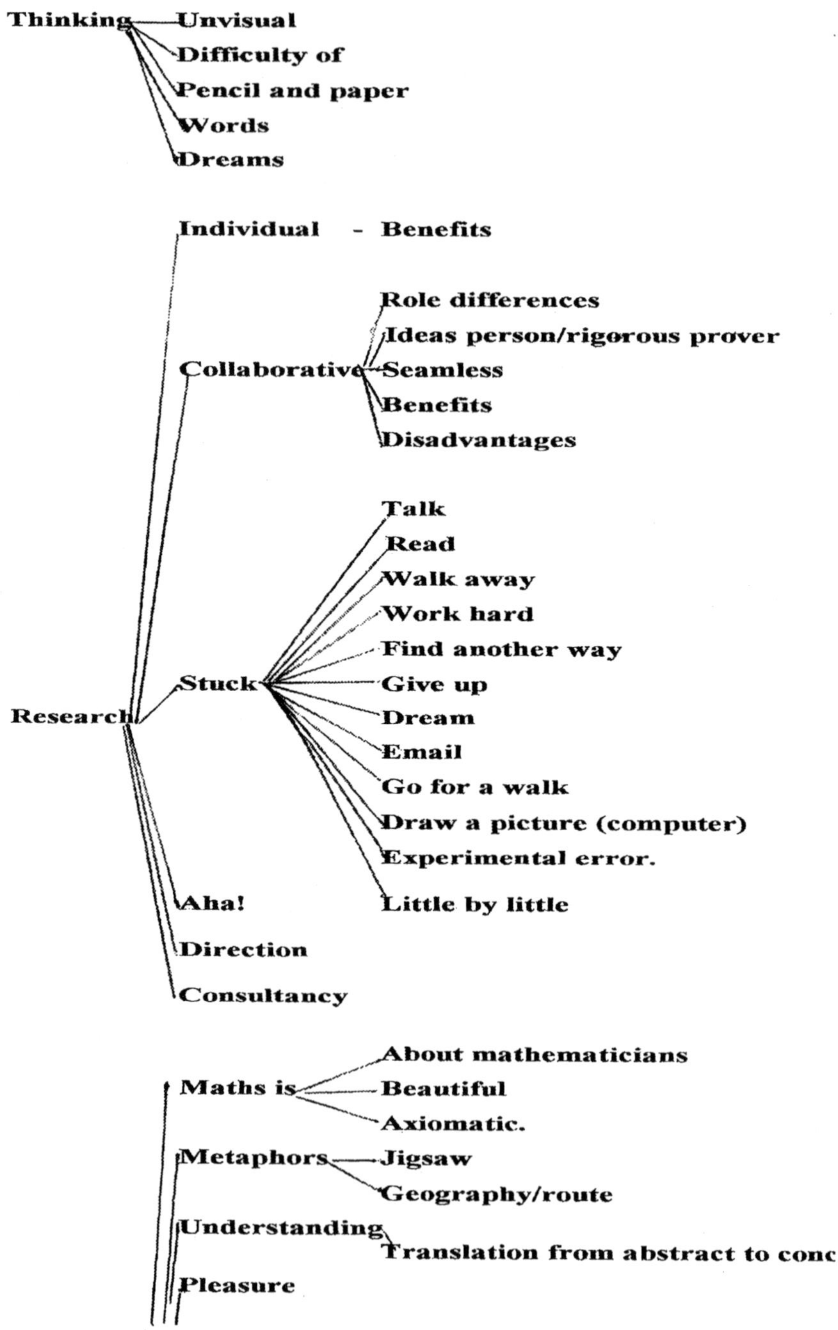
Thinking
Unvisual
Difficulty of
Pencil and paper
Words
Dreams
Individual - Benefits
Role differences
Ideas person/rigorous prover
Collaborative
Seamless
Benefits
Disadvantages
Talk
Read
Walk away
Work hard
Find another way
Stuck
Give up
Research
Dream
Email
Go for a walk
Draw a picture (computer)
Experimental error.
Aha!
Little by little
Direction
Consultancy
About mathematicians
Maths is
Beautiful
Axiomatic.
Metaphors
Jigsaw
Geography/route
Understanding
Translation from abstract to conc
Pleasure

- Writing style?
- Moving on or killing off a theory
- Personal comment
- Quality
- Sloppiness
- Check
- Fashionable

Feelings
- Good
  - Celebrate
  - Enjoy playing
  - Surprise
- Who are mathematicians?
- Bad - University
- Who achieves?
- About other mathematicians
- Confidence
- Excited - Not excited
- Computer role
- Oxbridge
- Completion
- Hard to write up

Problems
- Read
- Hear
- Number
- Talking
- Writing
- Continuation
- Brought by others
- Source of

Culture
- Competitive
- Collaborative
- Research
- Proof

REFERENCES

Alibert, D., & Thomas, M. (1991). Research on mathematical proof. In D. Tall (Ed.), *Advanced mathematical thinking* (pp. 215-230). Dordrecht, The Netherlands: Kluwer Academic.

Anderson, R. (1988). Overwriting and other techniques for success with academic articles. In P. C. Robinson (Ed.), *Academic writing: process and product* (pp. 151-158). London: Modern English Publications (in association with the British Council).

Boaler, J. (1997). *Experiencing school mathematics: Teaching styles, sex and setting.* Buckingham, UK: Open University Press.

Burton, L. (1999a). Fables: the tortoise? The hare? The mathematically under-achieving male? *Gender and Education*_11(4).

Burton, L., (Ed.). (1999b). *Learning mathematics - from hierarchies to networks.* London: Falmer.

Burton, L. (1999c). The practices of mathematicians: What do they tell us about coming to know mathematics? *Educational Studies in Mathematics 37*, 121-143.

Clark, R., & Ivanic, R. (1997). *The politics of writing.* London: Routledge.

Connolly, P., & Vilardi, T. (Eds.). (1989). *Writing to learn mathematics and science.* New York: Teachers' College Press.

Countryman, J. (1992). *Writing to learn mathematics: Strategies that work, K-12.* Portsmouth, NH: Heinemann.

Cronin, B. (1984). *The citation process: The role and significance of citations in scientific communication.* London: Taylor Graham.

Davis, P. J., & Hersh, R. (1981). *The mathematical experience.* Harmondsworth: Penguin Books.

Dixon, J. (1987). The question of genres. In I. Reid (Ed.), *The place of genre in learning: current debates.* Geelong: Centre for Studies in Literary Education, Deakin University.

Edwards, D., & Mercer, N. (1987). *Common knowledge: The development of understanding in the classroom.* London: Methuen.

Fairclough, N. (Ed.). (1992). *Critical Language Awareness.* Harlow: Longmans.

Galbraith, D., & Rijlaarsdam, G. (1999). Effective strategies for the teaching and learning of writing. *Learning and Instruction, Special Issue*, 9(2), 93-108.

Gerofsky, S. (1996). A linguistic and narrative view of word problems in mathematics education. *For the Learning of Mathematics*, 16(2), 36-45.

Gillman, L. (1987). *Writing mathematics well: A manual for authors.* Mathematical Association of America.

Halliday, M. A. K.(1973). *Explorations in the functions of language.* London: Edward Arnold.

Halliday, M.A.K. (1985). *An introduction to functional grammar.* London: Routledge & Kegan Paul.

Hersh, R. (1989). A mathematician's perspective. In P. Connolly & T. Vilardi (Eds.), *Writing to learn mathematics and science* (pp. 290-292). New York: Teachers' College Press.

Hodge, R., & Kress, G. (1993). *Language as ideology.* London: Routledge & Kegan Paul, (2nd ed.).

Hutton, Jane L. (1995). Statistics is essential for professional ethics. *Journal of Applied Philosophy*, 12(3), 253-261.

Knuth, D. E., Larrabee, T., & Roberts, P. M. (1989). *Mathematical writing.* Mathematical Association of America.

Krantz, S.G. (1997). *A primer of mathematical writing: Being a disquisition on having your ideas recorded, typeset, published, read, and appreciated.* Providence, RI: American Mathematical Society.

Kress, G. (1989). *Linguistic processes in sociocultural practice.* Oxford: Oxford University Press, (2nd ed.).

Kress, G. (1993). Genre as social process. In B. Cope & M. Kalantzis (Eds.), *The Powers of Literacy: A Genre Approach to Teaching Writing*, (pp. 22-37), London: Falmer Press.

MacNamara, A., & Roper, T. (1992). Unrecorded, unobserved and suppressed attainment: Can our pupils do more than we know? *Mathematics in School*, 21(5), 12-13.

Marks, G., & Mousley, J. (1990). Mathematics, education and genre: Dare we make the process writing mistake again? *Language and Education*, 4(2), 117-135.

Mason, J. (with Burton, L., & Stacey, K.). (1982). *Thinking mathematically*, London: Addison-Wesley.

Master, P. (1991). Active verbs with inanimate subjects in scientific prose. *English for Specific Purposes*, 10(1), 15-33.

Mathematical Association. (1987). *Maths talk.* Cheltenham: Stanley Thornes.

McIntosh,M.E.(1991). No time for writing in your class? *Mathematics Teacher*, 84, 423-433.

McLaren, P. (1995). *Critical pedagogy and predatory culture*, London: Routledge.

Morgan, C. (1996). "The language of mathematics": Towards a critical analysis of mathematical text. *For the Learning of Mathematics*, 16(3), 2-10.

Morgan, C. (1998). *Writing mathematically: the discourse of investigation.* London: Falmer Press.

Mousley, J., & Marks, G.(1991). *Discourses in mathematics.* Geelong: Deakin University.

National Council of Teachers of Mathematics. (1989). *Curriculum and evaluation standards for school mathematics.* Reston, VA: Author.

Pimm, D. (1984). Who is we? *Mathematics Teaching*, 107, 39-42.

Pimm, D. (1987). *Speaking mathematically: Communication in mathematics classrooms*, London: Routledge, Kegan & Paul.

Reid, I. (1987). *The place of genre in learning: Current debates.* Geelong: Centre for Studies in Literary Education, Deakin University.

Resnick, L. B., Levine, J. M., & Teasley, S. D. (Eds.) (1991). *Perspectives on socially shared cognition.* Washington, DC: American Psychological Association.

Richards,J. (1991). Mathematical discussions. In E. Von Glasersfeld (Ed.) *Radical constructivism in mathematics education* (pp. 13-51). Dordrecht: Kluwer Academic Publishers.

Rotman, B. (1988). Towards a semiotics of mathematics. *Semiotica*, 72(1/2), 1-35.

Rowland, T. (1995). Hedges in mathematical talk: linguistic pointers to uncertainty. *Educational Studies in Mathematics*, 29(4), 327-353.

Saussure, F. De. (1974). *Course in general linguistics* (J. Culler Ed., W. Baskin, Trans.). London: Fontana.

Secada, W.G., Fennema, E., & Adajian, L.B. (Eds.). (1995). *New directions for equity in mathematics education.* Cambridge, UK: Cambridge University Press.

Solomon, Y., & O'Neill, J. (1998). Mathematics and narrative. *Language and Education*, 12(3), 210-221.

Steenrod, N., Halmos, P. R., Schiffer, M. M., & Dieudonné, J. A. (1973). *How to write mathematics.* American Mathematical Society.

Steffe, L., & Gale, J. (Eds.). (1995) *Constructivism in education*, Hove, UK: Erlbaum.

Stephens, M., Waywood, A., Clarke, D., & Izard, J. (Eds.). (1993). *Communicating mathematics: Perspectives from classroom practice and current research.* Hawthorn, Vic.: Australian Council for Educational Research.

Ward, S. C. (1996). *Reconfiguring truth*, London: Rowman & Littlefield Publishers Inc.

INDEX

Mathematics Education Library

Managing Editor: A.J. Bishop, Melbourne, Australia

1. H. Freudenthal: *Didactical Phenomenology of Mathematical Structures.* 1983 ISBN 90-277-1535-1; Pb 90-277-2261-7
2. B. Christiansen, A. G. Howson and M. Otte (eds.): Perspectives on Mathematics Education. Papers submitted by Members of the Bacomet Group. 1986 ISBN 90-277-1929-2; Pb 90-277-2118-1
3. A. Treffers: Three Dimensions. A Model of Goal and Theory Description in Mathematics Instruction TheWiskobas Project. 1987 ISBN 90-277-2165-3
4. S. Mellin-Olsen: The Politics of Mathematics Education. 1987 ISBN 90-277-2350-8
5. E. Fischbein: Intuition in Science and Mathematics. An Educational Approach. 1987 ISBN 90-277-2506-3
6. A.J. Bishop: Mathematical Enculturation. A Cultural Perspective on Mathematics Education. 1988 ISBN 90-277-2646-9; Pb (1991) 0-7923-1270-8
7. E. von Glasersfeld (ed.): Radical Constructivism in Mathematics Education. 1991 ISBN 0-7923-1257-0
8. L. Streefland: Fractions in Realistic Mathematics Education. A Paradigm of Developmental Research. 1991 ISBN 0-7923-1282-1
9. H. Freudenthal: Revisiting Mathematics Education. China Lectures. 1991 ISBN 0-7923-1299-6
10. A.J. Bishop, S. Mellin-Olsen and J. van Dormolen (eds.): Mathematical Knowledge: Its Growth Through Teaching. 1991 ISBN 0-7923-1344-5
11. D. Tall (ed.): Advanced Mathematical Thinking. 1991 ISBN 0-7923-1456-5
12. R. Kapadia and M. Borovcnik (eds.): Chance Encounters: Probability in Education. 1991 ISBN 0-7923-1474-3
13. R. Biehler, R.W. Scholz, R. Str¨aßer and B. Winkelmann (eds.): Didactics of Mathematics as a Scientific Discipline. 1994 ISBN 0-7923-2613-X
14. S. Lerman (ed.): Cultural Perspectives on the Mathematics Classroom. 1994 ISBN 0-7923-2931-7
15. O. Skovsmose: Towards a Philosophy of Critical Mathematics Education. 1994 ISBN 0-7923-2932-5
16. H. Mansfield, N.A. Pateman and N. Bednarz (eds.): Mathematics for Tomorrow's Young Children. International Perspectives on Curriculum. 1996 ISBN 0-7923-3998-3
17. R. Noss and C. Hoyles: Windows on MathematicalMeanings. Learning Cultures and Computers. 1996 ISBN 0-7923-4073-6; Pb 0-7923-4074-4

Mathematics Education Library

18. N. Bednarz, C. Kieran and L. Lee (eds.): Approaches to Algebra. Perspectives for Research and Teaching. 1996 ISBN 0-7923-4145-7; Pb ISBN 0-7923-4168-6
19. G. Brousseau: Theory of Didactical Situations in Mathematics. Didactique des Math´ematiques 19701990. Edited and translated by N. Balacheff, M. Cooper, R. Sutherland and V. Warfield. 1997 ISBN 0-7923-4526-6
20. T. Brown: Mathematics Education and Language. Interpreting Hermeneutics and Post-Structuralism. 1997 ISBN 0-7923-4554-1
    Second Revised Edition. 2001 Pb ISBN 0-7923-6969-6
21. D. Coben, J. O'Donoghue and G.E. FitzSimons (eds.): Perspectives on Adults Learning Mathematics. Research and Practice. 2000 ISBN 0-7923-6415-5
22. R. Sutherland, T. Rojano, A. Bell and R. Lins (eds.): Perspectives on School Algebra. 2000 ISBN 0-7923-6462-7
23. J.-L. Dorier (ed.): On the Teaching of Linear Algebra. 2000 ISBN 0-7923-6539-9
24. A. Bessot and J. Ridgway (eds.): Education for Mathematics in the Workplace. 2000 ISBN 0-7923-6663-8
25. D. Clarke (ed.): Perspectives on Practice and Meaning in Mathematics and Science Classrooms. 2001 ISBN 0-7923-6938-6; Pb ISBN 0-7923-6939-4
26. J. Adler: Teaching Mathematics in Multilingual Classrooms. 2001 ISBN 0-7923-7079-1; Pb ISBN 0-7923-7080-5
27. G. de Abreu, A.J. Bishop and N.C. Presmeg (eds.): Transitions Between Contexts of Mathematical Practices. 2001 ISBN 0-7923-7185-2
28. G.E. FitzSimons:What Counts asMathematics? Technologies of Power in Adult and Vocational Education. 2002 ISBN 1-4020-0668-3
29. H.Alrø andO. Skovsmose:Dialogue and Learning inMathematicsEducation. Intention, Reflection, Critique. 2002 ISBN 1-4020-0998-4
30. K. Gravemeijer, R. Lehrer, B. van Oers and L. Verschaffel (eds.): Symbolizing, Modeling and Tool Use in Mathematics Education. 2002 ISBN 1-4020-1032-X
31. G.C. Leder, E. Pehkonen and G. T¨orner (eds.): Beliefs: A Hidden Variable in Mathematics Education? 2002 ISBN 1-4020-1057-5; Pb ISBN 1-4020-1058-3
32. R. Vithal: In Search of a Pedagogy of Conflict and Dialogue for Mathematics Education. 2003 ISBN 1-4020-1504-6
33. H.W. Heymann: Why Teach Mathematics? A Focus on General Education. 2003 ISBN 1-4020-1786-3
34. L.Burton: Mathematicians as Enquirers: Learning about Learning Mathematics. 2004 ISBN 1-4020-7853-6

KLUWER ACADEMIC PUBLISHERS-BOSTON /NEW YORK / DORDRECHT /LONDON